CARING ABOUT HUNGER

Caring About Hunger
First published 2016
by Irene Publishing

ISBN **978-91-88061-15-7**

Irene Publishing
Sparsnäs 1010, 66891 Ed, Sweden
irene.publishing@gmail.com
www.irenepublishing.com

Copyright by George Kent and Irene Publishing

Photos are under Creative Commons License

Layout by Jørgen Johansen

CARING ABOUT HUNGER

George Kent

University of Hawai'i
University of Sydney
Saybrook University

Contents

List of Figures	10
Preface	11
Acknowledgments	15

Chapter One
THE HUNGER HOLOCAUST — 17
Malnutrition Numbers	17
Mortality Numbers	17
Explanations	19
Genocide	24
Criminality	26
The Human Right to Adequate Food	28
Dignity	30

Chapter Two
CARING — 37
Being Nice to Each Other	37
Caring Individuals	41
Human Relations	42
Quantifying Caring	46
Caring in Communities	48
Animals and Plants	50
Artifacts and Nature	53
Past and Future, and Spirit	56
Evolutionary Biology	58
Increasing the Caring	63

Chapter Three
MOTIVATIONS FOR PRODUCING FOOD 69
Health or Wealth? 69
Agroecology 70
Pre-Modern Agriculture Today 71
Distance from Producers to Consumers 71
Distorted Value Chains 72
Solutions Without Problems 73
Linkage Between Agriculture and Nutrition 75
Marginalization of Food Workers 77

Chapter Four
WIDENING GAPS 81
Evidence 81
Explanations 83
Unequal Opportunities 85
Technological Advances 85

Chapter Five
CARING ABOUT CHILDREN 87
Children's Malnutrition 87
Commercial Baby Food 88
Human Milk Banking and Sharing 91
Responsibility and Care 93
Optional Protocol on Children's Nutrition 98

Chapter Six
GLOBAL AND NATIONAL ACTORS 101
Global Actors 101
Modest Aspirations 102
Long Distance Caring 106
Alternatives to Interventionism 109

Chapter Seven
DESIGNING CARING COMMUNITIES   113
The Concept   114
Social Arrangements   118
Physical Arrangements   121
Financial Arrangements   122
Economics   127
The Design Process   128

Chapter Eight
CARING ABOUT LOCAL HUNGER   129
Change From the Top?   129
Community-based Approach   130
Food Projects   133
Technology   137
Community Self-reliance   138
Wealth of Strong Communities   140

REFERENCES   144
Praise for *Caring About Hunger*   192
About the Author   199

LIST OF FIGURES

2-1	Caring: Mother and child	41
2-2.	Caring: Michelangelo's Piéta	42
2-3	Indifference: Hungry child	44
2-4.	Exploitation	45
2-5	Caring in the world of animals	52
5-1	Rings of responsibility	96

I want to express my gratitude to the photographers and artists who produced the figures used in this book. Most, found on the Internet, have unknown origins. I prepared the one on Rings of Responsibility for my 2005 book, *Freedom from Want: The Human Right to Adequate Food.*

Preface

I was close to finishing the manuscript for this book when I discovered an old note I had written to myself: "We already have all the technologies we need to produce all the food that is needed to feed everyone on earth very well, but it is not being used for that purpose. The motivation is not there. Thus, what is needed is not new technological inventions, but new social inventions, designed to harness the right kind of motivations." I forgot I had written that note, but the thought has gnawed at me all these years. The result is this book.

Why is there is so much hunger in the world? After all, the earth's resources are abundant. My first major effort to address the question resulted in a book called *The Political Economy of Hunger: The Silent Holocaust,* published in 1984. After struggling with the issue for a few more decades, I wrote an essay on *Designing a World Without Hunger* to, "stimulate discussion about what policies might be undertaken by overarching levels of governance to protect and facilitate local self-reliance in well-functioning local communities (Kent 2008a, 2)."

It took me a few more decades to recognize that while agencies at the global level and those in national governments do many good things, they really don't care enough about the hunger problem to do what needs to be done to solve it. They care, but not enough. They have other priorities.

We know there is widespread hunger in the world, but most of us set the thought aside. We know the problem is serious, but few of us could say how serious. It is distant, and we all have other important things to attend to, like children and jobs. Anyway, it is hard to see what we can do about the hunger problem, apart from writing a few checks for charities. But on second thought, maybe those charities should not be trusted. Forget those checks.

There is no technical obstacle to ending hunger. Despite the misleading accounts offered by some analysts and some food companies, there is no shortage of food in the world. The problem is that in our dominant food system, most food moves to people who have money, not to people who have serious unmet needs for food. The key missing element is caring.

We need better analyses of the social and political dimensions of the problem. Human relations such as caring, indifference, and exploitation matter a great deal. Yet the literature on the hunger issue is silent on these dimensions. Hunger is consistently presented as a technical problem, not a human relations problem. No wonder the problem has not been solved.

Progress is being made in addressing the hunger problem, but far too slowly. I now see that it is a mistake to always look upward to higher levels of governance for the answers. We should look inward, at local communities where people really care about one another's well-being, and see what can be done, not *for* but *by* the people there. "We" here means us outsiders working together with the insiders of the communities. Insiders are likely to care more than any outsiders about the well-being of the people there.

Chapter One takes a hard look at the issue of hunger and its magnitude. There are not many other things that cause so much sustained human misery. The magnitude of the harm it causes is amazing, and so is the fact that so few people know it.

Chapter Two explores the many dimensions of caring that hold societies together and sustain them over time. While this book is on caring about hunger, this chapter is about the idea of caring itself. It provides the foundation for the analysis in the subsequent chapters.

Chapter Three reflects on the major motivations for producing food. It argues that the historic switch from producing food mainly for the health of consumers to producing food mainly for the wealth of producers has messed up our world in many ways.

Chapter Four shows that the pursuit of unlimited wealth and power accelerates the widening of gaps between rich and poor. It fuels exploitation of the weak.

Chapter Five is on the widespread malnutrition among children. Many could be saved with better breastfeeding practices. Young children don't eat much, so it cannot cost much to feed them. The chapter focuses on the importance of having more distant actors care *about* children, and thus be more willing to support those at the front lines in their direct caring *for* children.

This book's main concern is what local communities could do to prevent hunger or respond to it when it occurs. Chapter Six, however, considers the role of higher level agencies. It recommends that instead of trying to address the hunger problem directly, distant agencies should work with local communities to increase their capacity to deal with hunger and related issues themselves.

Chapters Seven and Eight explore the concept of *caring communities*, where people really care about one another's well-being. Chapter Seven highlights the importance of the social dimensions in the design of caring communities. Caring can be strengthened by having people spend more time working and playing together.

Chapter Eight shows how that can be done while strengthening local food systems. It shows that in strong communities, where people really care about one another's well-being, no one goes hungry. That does not hold in cataclysmic disasters, but even then, no one goes hungry unless everyone goes hungry. If strong caring communities provide good models of how to live, others will adapt their ways of living to fit their own circumstances. In time, the process of ending hunger through the formation of such communities, one after the other, could lead to ending hunger globally.

There are no technical obstacles to ending hunger in the world. It depends on how people relate to one another, on dimensions such as caring, indifference, and exploitation. There is no easy way to measure caring, and no easy way to increase it, but at the very least there is a need to examine the role of caring in relation to hunger.

The idea of overcoming the forces that create and sustain hunger might sound utopian, but surely there is a need for a good understanding of the roots of the problem. The goal of ending hunger is distant but not unreachable. It has been achieved in many places, including places where people have little money. This approach to dealing with the issue gives us a better sense of the direction we need to take to move toward a world without hunger. That will be a world in which we have learned how to live together well.

Acknowledgments

Thank you, first, to Joan, our two young men, their wives, and the two wonderful granddaughters they gave us. All of them helped me understand the meaning of deep caring. My thanks also go to my relatives and ancestors, known and unknown, who set my path.

This book was built on foundations shaped by many different people and agencies. It was in the 1980s when, with the support of David James, I served as a consultant for the Food and Agriculture Organization of the United Nations. It was then that I learned that modern food producers generally don't care much about who gets to eat their food. They care mainly about sales, not nutrition.

I gave an invited talk at Webster University in St. Louis in 2010 on the Hunger Holocaust, published later that year. It became the basis for this book's opening chapter.

I wrote a chapter, "On Caring," for Michelle Brenner's edited volume, *Conversations on Compassion*, published in 2015. That chapter focused on caring itself, not food. Later I brought the two themes together in talks I gave on Caring About Hunger, first in Canberra, organized primarily by Ronald Johnson, and then in Sydney, organized primarily by Michelle Brenner. Through this transformation in Australia, the chapter "On Caring" in Michelle's book became the basis for Chapter Two of this book.

I soon realized that, given its importance, the theme of those talks, caring about hunger, should not be allowed to fade. That seeded the concept of this book.

I have had the good fortune to know and be influenced by many visionary leaders in global efforts to address hunger and other forms of malnutrition. I miss Urban Jonsson, who first drew me into global nutrition conferences. I have been mentored by many others, including Wenche Barth Eide, Asbjørn Eide, Geoffrey Cannon, and Ted Greiner. Pamela Morrison and James Akré have always been there to guide me on infant feeding issues.

In all of this, I appreciate the support of my employers. I served as a professor of political science at the University of Hawai'i for forty years, retiring in 2010. There I had the freedom to pursue my concerns, including the development of an online course on the Human Right to Adequate Food. I have been offering the same course since "retirement" as an Adjunct Professor at the University of Sydney in Australia and also Saybrook University in California. Both Jake Lynch at Sydney and Joel Federman at Saybrook gave me the space to develop my thinking on food issues.

Finally, I want to express my appreciation to Jørgen Johansen and Irene Publishing for their enthusiastic support for publication of this book. In the broad context of peace studies, they fully recognize the importance of structural violence and the need to address it through means that are empowering for its victims.

I am grateful for all these people and agencies, and many more, for all they have done to help make this project possible.

Chapter One

The Hunger Holocaust

MALNUTRITION NUMBERS

The annual reports on *The State of Food Insecurity in the World* show that close to a billion people do not get enough basic food to live an active life (FAO, IFAD, and WFP 2015). These reports focus on people who do not get enough food energy (calories) and counts them as *hungry*. The numbers vary from year to year, but they remain close to a billion.

We are concerned here not only with people who meet that report's narrow technical definition of hunger, but also the millions more who are seriously deficient in particular nutrients such as iron, iodine, and vitamin A, and those who suffer from weakened immune systems due to their poor diets. In addition, there are many people who are overweight and as a result are vulnerable to a variety of diseases. If we count the people who suffer from malnutrition in all its forms, the number is closer to two billion.

MORTALITY NUMBERS

The number of children dying before their fifth birthdays has been declining steadily, but according to the World Health Organization, "5.9 million children under age five died in 2015, 16 000 every day" (World Health Organization 2016b). It has been estimated, "undernutrition in the aggregate—including fetal growth restriction, stunting, wasting, and deficiencies of vitamin A and zinc along with suboptimum breastfeeding—is a cause of 3.1 million child deaths annually or 45% of all child deaths in 2011" (Black et al. 2013). The careful phrasing, "a cause", means that undernutriton is one among several causes contributing to child deaths. These are not all deaths due to outright starvation.

The international agencies focus on child malnutrition in low- and middle-income countries, but there is child malnutrition in high-income countries as well. It has been estimated that promoting breastfeeding "has the potential to save or delay ~720 postneonatal deaths in the United States each year" (Chen 2003, e435). Another study estimated, "If 90% of US families could comply with medical recommendations to breastfeed exclusively for 6 months, the United States would . . . prevent an excess 911 deaths, nearly all of which would be in infants . . ." (Bartick and Reinhold 2010). In high-income countries, there is far less mortality linked to malnutrition than in low-income countries, but there is no doubt that malnutrition results in serious harms to health for both children and adults in high as well as low income countries.

No national or global agencies regularly estimate how many deaths are related to malnutrition. Apart from the difficulty of making the estimates, it may be that this number is not estimated because people just don't care much about it.

Focus for a moment on the three million malnutrition-related deaths of children each year. It is important to get a sense of scale on this. Three million child deaths associated with malnutrition is far more than the total number of deaths (adults and children) due to armed conflicts. In 2013 the twenty deadliest wars in the world resulted in an estimated 127,134 deaths (PS21 Project 2015; World Bank 2015a). Cumulatively, United States war deaths since 1775 amount to about one million (Thompson 2016; also see DeBruyne and Leland 2015; Roser 2016b). The world's mass media give far more attention to the thousands of deaths in armed conflict each year than to the millions of deaths of children.

In a study of mortality associated with the war on terrorism, it was estimated that over a period of twelve years, the wars in Iraq, Afghanistan, and Pakistan resulted in about 1.3 million killed, directly or indirectly (International Physicians for the Prevention of Nuclear

War 2016). The report was discussed extensively in the media, where it was understood as showing the awfulness of those wars, also widely covered in the media. No one thought to compare the 1.3 million killed over twelve years as a result of direct violence with the six million deaths of children in one year as a result of structural violence (Kent 2006a), almost half of them associated with malnutrition.

The millions of nutrition-related deaths worldwide and the associated illnesses, repeated year after year, should be viewed as a type of holocaust. The phenomenon is different from the well-known deliberate genocides, but that difference does not make it unimportant. Why does widespread hunger and malnutrition happen? Why is it tolerated?

EXPLANATIONS

The scope of malnutrition worldwide is stunning not only because the numbers are so large, but also because so few people know about it. There are no serious technical obstacles to producing good food. The failure to decisively address the world hunger problem is largely the result of indifference, not caring enough. Most people have little interest in it.

People who are rich and powerful are not highly motivated to alleviate hunger and the poverty that underlies it. The consequences were captured in the title of a 2004 study, *Fatal Indifference: The G8, Africa and Global Health* (Labonte 2004). Its chapter on "Nutrition, Food Security and Biotechnology" concludes with simple understatement: "The lack of explicit commitments, goals and strategies related to enhanced food security, especially in the regions of the world where undernourishment is most prevalent, is disturbing."

Some individuals and nongovernmental organizations care deeply, but the motivation to end hunger is not at the level it needs to be. Hunger could be ended quickly if the people who have the power cared enough about the people who have the problem. When she was U.S. Secretary of State, Hillary Clinton got it right when she said, "the question is not whether we can end hunger, it's whether we will" (On the Hill 2009; also see Butterly and Shepherd 2010). There are no major technical obstacles.

There is widespread indifference. There is also the fact that many people benefit from the existence of hunger. Whether consciously or unconsciously, many people and many countries act in ways that ensure that hunger in the world is sustained.

We sometime see poor people by the roadside holding up signs that say, "Will Work for Food." Most people work mainly for food. People need food to survive, so they work, either in producing food for themselves in subsistence-level production, or by selling their services to others in exchange for money. Hunger is fundamental to the functioning of the world's economy.

Hungry people are productive workers, especially where there is a need for manual labor. They are the front-line producers and processors in food production and in many other industries. Workers' productivity is not indicated by the level of their earnings, but many companies make major contributions to the gross national product by using poorly paid workers. Those who have few options sell their services cheaply. In doing that the poor help to enrich those who own the factories and the machines and the land, and help to provide consumers with inexpensive products.

For employers who depend on cheap labor, hunger is the foundation of their wealth. In his *Dissertation on the Poor Laws in England* of 1786, Joseph Townsend observed, "... hunger is not only a peaceable, silent, unremitted pressure, but, as the most natural motive to industry and labour, it calls forth the most powerful exertions . . ." (Townsend 1786).

The famed essayist Ralph Waldo Emerson saw that people often pay for their food with their independence:

> *Society everywhere is in conspiracy against the manhood of every one of its members. Society is a joint-stock company, in which the members agree, for the better securing of his bread to each shareholder, to surrender the liberty and culture of the eater. The virtue in most request is conformity. Self-reliance is its aversion. (Emerson 1841)*

Usually hunger motivates people to work cheaply in quiet ways, but at times it is blatant, as in the case of "Tetley's Tata Tea Starving Indian Tea Workers into Submission":

> *Tata, the transnational Indian conglomerate whose Tetley Group makes the world famous Tetley teas, has taken 6,500 people hostage through hunger. The hostages are nearly 1,000 tea plantation workers and their families on the Nowera Nuddy Tea Estate in West Bengal, India. Permanently living on the edge of hunger, the workers and their dependents are being pushed to the edge of starvation through an extended lock out which has deprived them of wages for all but two days since the beginning of August. The goal of this collective punishment is to starve the workers into renouncing their elementary human rights, including the right to protest extreme abuse and exploitation. . . . (IUF 2009)*

The story of these tea workers continues as a story of life without dignity (FIAN International 2016).

The nutrition literature says it is important to ensure that people are well fed so they can be more productive. That is misleading. No one works harder than hungry people. People who are well nourished have greater *capacity* for productive physical activity, but they are far less *willing* to do that work.

If there were no hunger in the world, who would do all the hard work in fields and factories? Hunger ensures that many people will work cheaply. That is a blessing to those who benefit from the fruits of their labor, whether as employers or as consumers. People at the high end are not rushing to solve the hunger problem because for them hunger is an asset. For those who depend on the availability of cheap labor, poverty is the foundation of their wealth.

Low-paying jobs cause hunger. For example, one report told about "Brazil's ethanol slaves: 200,000 migrant sugar cutters who prop up renewable energy boom" (Phillips 2007; also see Simoes 2008). While it is true that low-paying jobs create hunger, at the same time hunger causes the creation of low-paying jobs. No one would establish massive biofuel production operations in Brazil if they did not know there were thousands of hungry people ready to take the awful jobs they offer. No one would build a factory if they did not know people would be available to take the jobs. We would not have fast food restaurants if the owners did not know there were many people who would work for low wages.

A nongovernmental organization, Free the Slaves, estimates there are somewhere between 21 and 36 million slaves in the world (Free the Slaves 2016). They define slaves as people who are not allowed to walk away from their jobs. While there are many people who are literally locked into workrooms, others work in slave-like conditions, such as bonded laborers in south Asia and migrant farm workers in the United States (Dooley 2009; Immokalee 2008). Apparently, the count by Free the Slaves does not include those who might be described as slaves to hunger, people who are free to walk away from their jobs, but have nothing better to go to. Many people are slaves to hunger.

Hunger is allowed to persist partly because it is believed that there is a need for low wage workers to ensure good incomes to employers and affordable products to consumers. Several other explanations have been offered as well. Many people feel that ending hunger through assistance program would be too costly, and there are many other ways in which the resources could be used. Some believe that generous social welfare programs would produce waves of unwanted in-migration. Some believe that hunger helps to protect the world from runaway population growth.

We sometimes talk about hunger and poverty as scourges that all of us want to see abolished, but that naïve view prevents us from coming to grips with what causes and sustains them. Hunger and poverty have positive value to many people.

Why end malnutrition? Most people will want to prevent or remedy their own serious malnutrition or that of their family members. The more difficult question is, why should any of us want to deal with the malnutrition problems of others, especially distant others who we don't know and will never meet? We need to have a clear answer to the *why* question if are ever to have good answers to the *how* question.

There are often grand efforts to provide assistance during sudden-onset disasters like the earthquake in Haiti in January 2010. But what was the world's response to the preceding decades of chronic hunger in Haiti, and in the years since the 2010 earthquake? Given the persistence of hunger and poverty in the world, we have to wonder whether the ongoing challenge of hunger has ever been addressed in a serious way.

Despite the comforting rhetoric, there is not a harmony of interests between those who suffer and those who have the power to solve those problems. Hunger and poverty are manifestations of structural violence that is mediated through social systems. This structural violence leads to the endless re-creation of hunger and poverty. Summarizing, hunger and poverty persist because of the powerlessness of the poor and the indifference and exploitativeness of the rich. There are three key points:

- Disjunction. Hunger and poverty persist largely because the people who have the power to solve the problems are not the ones who have the problems.

- Compassion. On the whole, the people who have the power do not have much compassion for the powerless.

- Material interests. The powerful serve mainly the powerful, not the powerless, because the powerless cannot do much for the benefit of the powerful. In many cases the powerful exploit the powerless.

While we may wish it were otherwise, it is factually a mistake to view humanity as one. We are divided, so it matters that the costs of ending hunger and poverty would go to one group while the benefits would go to another. Hunger persists because the powerful have the capacity but not the will to address the problems adequately, while the powerless have the will but not the capacity.

The disjunction between those who have the problems and those who have the power to solve them is seen both within countries and globally. The lack of caring within countries is evident when we see widespread malnutrition side by side with concentrated wealth, a theme highlighted in Chapter Four, on Widening Gaps.

GENOCIDE

On June 24, 1981 a group of 52 Nobel Prize laureates issued a *Manifesto Against Hunger*:

> We ... address an appeal to all men and women of good will ... so that dozens of millions of those who are suffering from starvation and underdevelopment, victims of the international political and economic disorder so widespread today, may be restored to life.
>
> An unprecedented holocaust, whose horror includes in a single year all the horror of the exterminations which our generations saw in the first half of the century, is still happening today and continuing to widen, every moment that passes, the perimeter of barbarities and death in the world, no less than in our consciences.
>
> All those who have taken stock of the holocaust, who are publicising it and fighting it, are unanimous in defining politics first and foremost as the cause of this tragedy ... *(Manifesto 1981)*.

There is room for debate about whether hunger worldwide, allowing the deaths of millions of people each year, amounts to a form of genocide. The *Convention on the Prevention and Punishment of the Crime of Genocide* was adopted by the United Nations General Assembly on December 9, 1948 and entered into force on January 12, 1951. According to article II:

> In the present Convention, genocide means any of the following acts committed with intent to destroy, in whole or in part, a national, ethnical, racial or religious group as such:

(a) Killing members of the group;

(b) Causing serious bodily or mental harm to members of the group;

(c) Deliberately inflicting on the group conditions of life calculated to bring about its physical destruction in whole or in part;

(d) Imposing measures intended to prevent births within the group;

(e) Forcibly transferring children of the group to another group. *(UNOHCHR 2010a)*

The hungry do not constitute a group in the sense indicated in the convention. It recognizes only national, ethnic, racial, and religious groups as potential victims of genocide. Thus, in terms of this convention, the idea of calling hunger genocidal is questionable.

The convention also specifies there must be the *intent* to destroy if an action is to be identified as genocide. However, hunger does not result from the deliberate action of readily identified actors in the pattern characteristic of other commonly recognized genocides.

Deliberate neglect describes the pattern of many governments' responses to hunger. It is comparable to the disputed concept in law of "willful negligence" or "advertent negligence," defined as "the type of negligence that is deliberate with the intentional disregard for other people's welfare" (The Law Dictionary 2016).

The term is not self-contradictory. Neglect can be understood as the failure to do something that should be done—and that failure may or may not be intentional. If it persists and it is obvious, it must be regarded as intentional. If the failure to attend to people's needs persists over time, even in the face of repeated complaints and appeals, that neglect should be described as deliberate.

There is a difference between not knowing what your actions will lead to and what is described in law as "reckless disregard" for the predictable consequences of one's action. Manufacturers of cars and pharmaceuticals are expected to pull their products off the market if they learn they have serious harmful effects. Ignoring that harm while having full knowledge of it is a crime.

When infant formula was first promoted in low-income countries, it might not have been anticipated that it would kill babies. But when international governmental and nongovernmental organizations documented and warned and campaigned about the problem, and the World Health Assembly passed guidelines to control the behavior of sellers of infant formula, and still the sellers persisted in selling the product in a way that is known to kill babies, that is unforgivable. It is a form of killing. Those who are interested have been aware of this problem since Dr. Cicely Williams gave her "Milk and Murder" speech in 1939 (Baby Milk Action 2010).

In other genocides, killings are concentrated in a particular time and space. However, deaths from hunger are dispersed all over the globe and they are sustained over time. There is no central command structure causing these deaths to happen. There is nothing like the Wannsee conference of January 20, 1942 at which the Nazis systematically set out their plans for the extermination of the Jews of Europe.

There is that difference. The widespread neglect of hunger is not the calculated program of a few madmen assembled at a particular moment in history. The massive mortality due to hunger is more frightening precisely because it occurs worldwide with no central coordination mechanism. The culpability is not individual but systemic.

CRIMINALITY

Deaths from malnutrition are not described as murders, but that does not mean they are accidental or natural or inevitable. Some are self-inflicted, resulting from unwise personal dietary choices. The vast majority can be described as resulting from a form of negligent homicide by the surrounding society. Negligent homicide is still homicide.

Some argue that genocide should be defined narrowly, as it is in the genocide convention, to prevent the debasement of the concept. But there is no reason to suggest that other kinds of avoidable large-scale mortality are less important. A sensible alternative would be to acknowledge that there are different *kinds* of genocide associated with different categories of victims and different forms of intentionality.

This is the approach advocated by Israel Charny in his taxonomic scheme. He defines genocide in the generic sense as the willful destruction of a large number of human beings, except as that might be necessary in self defense. He then suggests that in distinguishing different categories of genocide, the degree of willfulness or intentionality should be assessed, leading to rating of different degrees of the crime of genocide (Charny 1994).

Perhaps the definitions used in assessing homicides could be adapted. Just as there can be first, second, or third degree murder, so too there might be first, second, or third degree genocide. Further distinctions could be made to take account of sustained deliberate neglect.

Widespread hunger deaths differ in many ways from the Holocaust and other atrocities we commonly describe as genocides. The differences, however, are not sufficient to dismiss the issue. The conclusion is inescapable: *hunger is so massive, so persistent, and so unnecessary, allowing it to continue should be recognized as a kind of genocide.*

Hunger is not the particular type of genocide specified in the genocide convention, but as this large-scale ongoing deliberate neglect results in the avoidable deaths of many millions of people, it too should be viewed as a crime comparable to other types of genocide.

Families and countries may be poor, but the world is not poor. There are no technical mysteries about how to prevent hunger. Ongoing, widespread, intense malnutrition should be recognized as *prima facie* evidence of an ongoing crime by society. As a crime there should be mechanisms in law for correcting that manifest injustice, including means for calling not only parents and local communities but also governments to account. The foundation of that mechanism would be the clear recognition in law and practice of human rights.

THE HUMAN RIGHT TO ADEQUATE FOOD

The right to life is well established in international human rights law, so on that basis alone, widespread hunger should be recognized as violating human rights. In India the entire right to food movement is based on the simple assertion of the right to life in the nation's constitution.

There is no need to derive arguments from the right to life alone. The human right to adequate food is now well developed in international human rights law (FAO 2016a; Kent 2005).

While the first phase of this effort focused on the obligations of national governments to people living under their jurisdiction, later work gave attention to the external obligations of states (Künnemann 2016) and the obligations of the global community taken as a whole (Kent 2008b). The right to food guidelines say that national development efforts should be supported by an enabling international environment. Relevant international agencies "are urged to take actions in supporting national development efforts for the progressive realization of the right to adequate food in the context of national food security" (FAO 2005, 33).

The core document of the modern global human rights system, the *Universal Declaration of Human Rights*, says in article 28:

> *Everyone is entitled to a social and international order in which the rights and freedoms set forth in this Declaration can be fully realized. (UNOHCHR 2010c)*

This in turn stands on the *United Nations Charter*, which says, in article 55:

> *With a view to the creation of conditions of stability and well-being which are necessary for peaceful and friendly relations among nations based on respect for the principle of equal rights and self-determination of peoples, the United Nations shall promote:*

> a. *higher standards of living, full employment, and conditions of economic and social progress and development;*
>
> b. *solutions of international economic, social, health, and related problems; and international cultural and educational cooperation; and*
>
> c. *universal respect for, and observance of, human rights and fundamental freedoms for all without distinction as to race, sex, language, or religion. (UNOHCHR 2010b)*

Article 56 of the Charter says:

> *All Members pledge themselves to take joint and separate action in co-operation with the Organization for the achievement of the purposes set forth in Article 55.*

Thus the charter and the declaration clearly acknowledge the responsibility of the global community, taken as a whole, for the realization of human rights. If everyone is entitled to an international order that will ensure the full realization of all human rights, we must work on envisioning and establishing such an order. Surely it should be an order in which the world as a whole recognizes not only moral responsibilities but also legal obligations for the realization of those rights. We must begin with the understanding that there are global obligations that are beyond those of states to their own people. Then we can begin to work out their exact content.

The human right to adequate food means there is an obligation to reach the goal of ending hunger and ensuring food security for all. These obligations fall primarily on national governments, but in some ways they are shared by all of us. There are choices that can be made with regard to means, but there is no choice with regard to the obligation to move decisively toward the goal. Thus, concrete obligations for ensuring realization of the human right to adequate food for all can be identified through the formulation of a concrete strategy for realizing that goal. Once one knows what steps are required to reach the goal, then there is an obligation to take those steps. If there are several different ways to reach the goal, choices may be made among them, but there is an obligation to choose some path that can realistically be expected to reach the goal.

There have been many global programs for responding to large-scale malnutrition, but they propose only to work around the edges of the problem, not to end it. There is a need for a global strategy and program of action that really could be expected to end hunger as a major public policy issue in the world. Not only moral considerations but also a fair interpretation of human rights law and principles require such a strategy and program of action.

Some children are born into poor countries, but no child is born into a poor world. Each child has rights claims not only against its own country and its own people; it also has claims against the entire world. If human rights are meaningful, they must be seen as universal, and not merely local. Neither rights nor obligations end at national borders. While national governments have primary responsibility for ensuring the realization of the human right to adequate food for people under their jurisdiction, all of us are responsible for all of us, in some measure. The task is to work out the nature and the depth of those global obligations.

DIGNITY

In 1999 the United Nations Committee on Economic, Social and Cultural Rights, which deals with the right to food, recognized the central importance of dignity in relation to the right to food:

> *The Committee affirms that the right to adequate food is indivisibly linked to the inherent dignity of the human person and is indispensable for the fulfilment of other human rights enshrined in the International Bill of Human Rights. It is also inseparable from social justice, requiring the adoption of appropriate economic, environmental and social policies, at both the national and international levels, oriented to the eradication of poverty and the fulfilment of all human rights for all. . . .*

The right to adequate food shall therefore not be interpreted in a narrow or restrictive sense which equates it with a minimum package of calories, proteins and other specific nutrients. (UNECOSOC 1999, paragraphs 4, 6)

This emphasis on dignity is grounded in the Universal Declaration of Human Rights of 1948, the document that launched the modern human rights system. It begins by saying, "recognition of the inherent dignity and of the equal and inalienable rights of all members of the human family is the foundation of freedom, justice and peace in the world." The first article affirms, "All human beings are born free and equal in dignity and rights."

The human right to food should not be understood simply as a matter of delivering nutrients to passive rights holders. As Ivan Illich put it, people need to provide for themselves because "people die when they are fed" (Illich 1973). While it is technically possible to assure that individuals' biological nutritional needs are fulfilled through tubes or by force, that would be different from fulfilling their human right to food. Serving pork to Muslim prisoners would violate their human rights, even if it contains all the nutrients they need. Dignity does not come from being fed. It comes from providing for oneself. In any well-structured society, the objective is to move toward conditions under which all people can provide for themselves (Kent 2005, 46).

People can suffer many indignities under uncaring implementation of legal rights. To take just one example, under India's version of the right to food, people are entitled to purchase highly subsidized grains at local ration shops. But for a village in Rajasthan, the ration shop, five kilometers away, is open only three days a month, and generally has long lines of people waiting for their share. Walking to the shop and then waiting, they lose one or more days of work and the income that comes with it. In one month that was monitored, sixty percent of the eligible families did not receive any rations. The families have no meaningful recourse, no way to challenge the system to get it corrected (Goel 2016).

This sort of thing happens over and over again because powerful people do not care enough about ordinary people's well-being. They are not going to care just because they have obligations under the law. They will accept such obligations and take them seriously if and when they care.

Delivering packaged meals in the way one might deliver feed pellets to livestock would be incompatible with human dignity. While delivering packaged meals might be sensible in a short-term emergency, it cannot be the means for realizing the human right to adequate food over the long run.

One of the major critiques of humanitarian assistance programs has been that "Aid processes treat lives to be saved as bare life, not as lives with a political voice (Edkins 2000, xvi)." If people are to be addressed as dignified human beings, they should have a say on how they are treated. This is why every human rights-based program should have safe and effective recourse mechanisms available to the rights holders themselves. People should have institutionalized remedies available to them that they can call upon if they feel they are not being treated properly.

Institutionalized recourse mechanisms ensure that rights holders have a voice and thus a measure of dignity. Human rights are not simply about setting standards. The core of any human rights system lies in the way in which it ensures rights holders will be heard. People must be able to claim their own rights, and not depend on others claiming their rights for them.

Concern for dignity also means that with regard to the levels of state obligations, high priority should be placed on government action that is *facilitating*, by establishing enabling conditions that allow people to provide for themselves. *Providing* food directly should take priority only when people cannot provide for themselves for reasons beyond their control. Human rights are mainly about upholding human dignity, not about meeting physiological needs. Dignity does not come from being fed. It comes from providing for oneself.

The objective should not be for governments to feed people, but to move toward conditions under which all people can provide for themselves. If people have no chance to influence what and how they are being fed, if they are fed prepackaged rations or capsules or are fed from a trough, their right to adequate food is not being met, even if they get all the nutrients their bodies need. The hunger problem needs to be handled not as one would approach livestock management but rather as a partnership, based on genuine concern for the well-being of those who are hungry, and with the direct engagement of the hungry in addressing the problem.

Rights should be based on clear recognition of and respect for human dignity. As Joel Feinberg said:

> *Having rights enables us to "stand up like men," to look others in the eye, and to feel in some fundamental way the equal of anyone. To think of oneself as the holder of rights is not to be unduly but properly proud, to have that minimal self-respect that is necessary to be worthy of the love and esteem of others. Indeed, respect for persons . . . may simply be respect for their rights, so that there cannot be the one without the other (Feinberg 1980, 151).*

Human rights are not a benevolent gift from elites, but the legal expression of every individual's entitlement to dignity.

In 2008 the United Nations General Assembly adopted a resolution that "would have the Assembly reaffirm that hunger constitutes an outrage and a violation of human dignity, requiring the adoption of urgent measures at the national, regional and international level, for its elimination" (United Nations General Assembly 2008). The United States was the only country to vote against it, going against the widespread recognition of the importance of dignity in relation to the right to food and to human rights generally. The United States has not ratified the Convention on Economic, Social and Cultural Rights, the human rights treaty that serves as the basis for the human right to adequate food. It is the only country that has not ratified the Convention on the Rights of the Child. The United States government consistently refuses to accept the application of the right to food for people under its jurisdiciton, and it does not accept that states have particular extraterritorial obligations arising from a right to food (International Human Rights Clinic 2013; Messer and Cohen 2007; Robi 2014).

The hunger problem is not only about the dignity of those who are hungry. It is about the dignity of all of us. When the United States repeatedly speaks out at the United Nations against the very idea of the human right to adequate food, I am embarrassed by my citizenship. When I see widespread misery around the world that is wholly unnecessary, I am ashamed, not for myself, not for the hungry, but for all of humankind.

People commonly ask how it will be possible to feed future generations. The question is insulting. Why ask how people are to be fed, as if this had to be done by some external agent? Most people are motivated to provide for themselves, and only need decent opportunities to do that. Who, when not deprived of the means, would not feed themselves and their families?

Worldwide, there is no shortage of food. There is a shortage of decent opportunities for poor people to produce food or to earn enough money to buy food for their families. The world's leaders only have to ensure that all people have decent opportunities to provide for themselves. Then the leaders just have to get out of the way, and let hunger end. It is not complicated, if the will is there.

In 1984 the editors of one of the earliest books on the human right to adequate food said:

> *At the conclusion of the World Food Conference held in Rome 1974 the governments of the world proclaimed "that within a decade no child will go to bed hungry, that no family will fear for its next day's bread, and that no human being's future and capacities will be stunted by malnutrition."*
>
> *As that decade comes to a close the tragic reality is that little, if any,*

progress has been made toward meeting those goals. During the target year of 1984, as during every other year since the Conference, literally millions of children have starved to death, tens of millions have gone to bed hungry and malnutrition continues to afflict hundreds of millions of people in all parts of the world. These statistics make hunger by far the most flagrant and widespread of all human rights abuses. (Alston 1984, 7)

Decades later their conclusion remains true. No violation of human rights has done more harm to more people than hunger. If no decisive action is taken, it will retain that distinction for years to come.

Without caring, human rights and social programs don't work well. We need to come to a better understanding of caring, and we also need to find ways to increase the caring. The challenge of ending hunger in a world of abundant resources dramatically demonstrates its importance.

Chapter Two

Caring

People who work on large-scale social problems become engrossed in the dark side of human behavior, studying things like violence, poverty, hunger, and other social pathologies. We begin to see these things as common, even inevitable. We look for remedies for these phenomena when they occur, and rarely consider that maybe they need not occur. Instead of just studying war and hunger, we should also give attention to how peace and plenty might work.

This chapter explores the meaning of caring in its many dimensions. The premise here is that caring about people, animals, things, the environment and much more is the glue that keeps us together and, at least potentially, can keep us from killing each other and the planet.

BEING NICE TO EACH OTHER

Most people most of the time treat each other rather nicely. Surely, caring must be a central element. We do that without fanfare. Caring is so common, we don't see it.

In *Mutual Aid*, Peter Kropotkin pointed out that people usually treat each other well, helping each other out in countless ways (Kropotkin 1902). Anthropologist Marshall Sahlins described how this worked in pre-modern societies (Sahlins 1972). Modern social scientists have little to say about this good quality in human relations. It is as if they have forgotten the *social* in social science. Social scientists should give more attention to the reality that people do a lot to take care of one another.

Thinking in terms of conventional economic analysis can make people less caring (Grant 2013a). That framework is driven by the idea of individual accumulation of wealth, but this is not always the dominant motivation. In reality, people share resources in many different ways, and many people refuse to be stunted by the compulsive desire for endless accumulation. There are some forms of economic analysis that do not make the wrong-headed assumptions we learn in Introductory Economics classes. Analyses of the *gift economy*, for example, are based on the understanding that people tend to be generous to one another. So far the alternative ways of thinking about economics have not displaced the dominant economic framework. It focuses on commercial transactions, not human relationships.

This chapter is about caring, defined as acting to benefit others. The term can refer to the action or to the underlying motivation for it. Our focus is on empathetic (or empathic) caring, the caring that is rooted in our capacity to share others' feelings. Empathetic caring is distinguished from instrumental caring of the sort done by a hired caretaker. Empathetic caring is about action taken *for the purpose* of benefiting others.

Empathy goes beyond mere cognitive understanding of how others feel to also include an emotional impact:

> *It is feeling sad in response to another's sadness; joy in response to another's joy; fear in response to another's fear, and so on. So conceived, empathy transfers others from external objects into parts of ourselves; "different" consciousnesses not only interact, they interpenetrate. In this way empathy expands our identity to include others; what happens to them, in some measure, happens to us. (Contri 2011)*

With caring, your well-being is linked to and affected by others' well-being. Their feeling good makes you feel good, and their feeling bad makes you feel bad.

Empathetic caring should be distinguished from instrumental caring, the kind that is offered in exchange for some direct benefit to oneself. Empathetic caring is its own reward. Caring that is primarily a means for obtaining benefits for the carer is instrumental caring (Grant 2013b).

In organized business-like caring, usually there is a clear distinction between those who have needs and those who provide caring for them, as in a care home. In natural caring, there is no structural distinction between those who provide care and those who receive it. With strong mutual caring, there is much less need for deliberately designed caring by specialists. Often, increasing needs for caring interventions by specialists are signs of the weakening of natural caring in the community.

At times instrumental caring becomes mechanistic, done by a "caretaker" mainly because he or she is paid to do it.

Instrumental caring is based on self-interest, while empathetic caring is based on concern for the well-being of another. Instrumental caring is not bad, but we should recognize that it is different from empathetic caring.

Often, caring acts are undertaken for mixed motives, partly to benefit another, and partly to benefit oneself. For example, the Fonterra dairy company in New Zealand explains that it provides free milk to school children "because we believe it will make a lasting difference to the health of New Zealand's children" (Fairfax NZ News 2012), but elsewhere the company "admits it's also about promoting its product, and lifting falling milk consumption" (3 News 2012). Apparently Fonterra distributes free milk to improve children's health and also to improve its long-term profits. There is nothing wrong with doing both at the same time. We often do things for several different reasons.

At times there are good reasons for cynicism. Along with promoting milk consumption at home and in other parts of the world, Fonterra is promoting the use of infant formula, especially in Asia. In that context, it does not mention that this profit-motivated push is likely to result in worse health outcomes for millions of infants (Kent 2011b, 2015c). In other contexts, we see manufacturers marketing foods to children with blatant disregard for their impact on children's health (Simon 2012). The widespread tendency of the food industry to prioritize their profits over consumers' health sends a message about their caring priorities (Mustain 2013).

Some writers define altruism as "behavior carried out to benefit another without anticipation of rewards from external sources" (Macaulay and Berkowitz 1970). That is described here as empathetic care. It does not require self-sacrifice. There is nothing intrinsically wrong with drawing benefits for oneself at the same time one acts to benefit others. Many people who are called to do good things for others enjoy their calling immensely.

Empathetic caring is usually a good thing. But there can be harm when empathy for one person or group leads to action that harms others. For example, armed conflict generally is undertaken to benefit a specific group at the expense of others.

Caring and compassion are closely related. Some analysts define compassion as "the feeling that arises in witnessing another's suffering and that motivates a subsequent desire to help" (Goetz, Keltner, and Simon-Thomas 2010). In the perspective adopted here, caring is not always triggered by suffering. The abundant caring behavior observable in ordinary daily life is not always in response to suffering.

Caring may be related to cooperation, but there can be cooperation without caring. Cooperation is often about making a deal based primarily on self-interest. Business people often cooperate because they feel they can make more profit by working together rather than separately.

At times cooperation turns into collusion or conspiracy, where there is cooperation at the expense of some third party.

Some writers focus on cooperation in situations in which individuals seek to maximize their self-interest (Axelrod 2006). One book devotes a section to the point that cooperation is not altruism (Ratner 2012). Cooperation that is driven primarily by self-interest is what we have described as instrumental caring. It is different from empathetic caring, which is about doing things for the purpose of benefiting others.

Empathetic caring does not have to be explained in terms of its producing material advantage for oneself or one's group. Indeed, the concept of deal-making demeans that type of caring. Mother Teresa and Saint Damien did not care for the homeless and the sick in exchange for an anticipated payoff.

Empathetic caring should be made more visible so its importance can be recognized. This chapter explores the idea of caring, beginning with the more familiar types, and then pushes the envelope to consider less familiar types. There is caring for one another, for animals, for things, for our ancestors, for the planet, for the cosmos. They are all somehow connected together, and that connectedness is both mysterious and important. This should help us get a deeper appreciation of how it is that we care and thus act to benefit other people and other things, and it might help us find ways to increase the caring.

CARING INDIVIDUALS

Caring can be a simple and beautiful thing, as illustrated here.

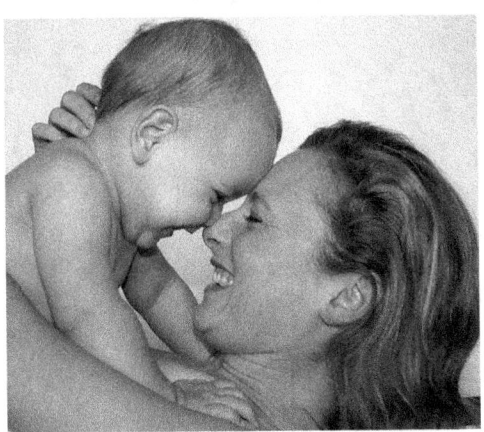

Figure 2-1. Caring: Mother and child

There are deeper multi-layered representations of caring, like that in Michelangelo's Piéta.

Figure 2-2. Caring: Michelangelo's Piéta

One doesn't need to know about Christianity or the fact that that statue has an honored place in Saint Peter's Basilica in Rome to see and feel that it is about caring. Michelangelo was so powerful as an artist that his portrayal transcends the imagery of two specific people to instead represent caring in all of humankind.

HUMAN RELATIONS

Simplifying, we can distinguish three major types of human relations:

- **Caring**, in which people feel better off when other individuals are better off, and thus act to benefit others;

- **Indifference**, in which people do not care about others' well-being, and thus do not act to help them or to harm them; and

- **Exploitation**, in which people benefit from others' misery.

An employer who pays his workers as little as possible is exploitative. A consumer who buys cheap products from ruthless manufacturers could be seen as an exploiter. Sadists get pleasure from others' misery. Germans have a special term, *schadenfreude*, for "enjoyment obtained from the troubles of others" (Merriam-Webster 2016). We can include in this category the pleasure a general might feel in defeating an enemy, or a coach might feel when his team defeats an opposing team. Rapists and pornographers are exploiters. Exploiters draw benefits from hurting others.

Exploitation is often done for what are claimed to be good reasons. You might hire low-wage workers for your business so that you can provide more comforts for your family. You might kill others to protect your country. You might accept collateral damage from a bombing run because you calculate that the benefit of killing militants outweighs that cost. You might discriminate against people of certain races in order to protect your own.

People weigh and balance things differently. We don't all navigate by the same moral compass. Whether intentionally or not, we often benefit from actions that harm others. Caring, indifference, and exploitation can be tied together in complex ways.

These are broad and crude categories, but they convey useful distinctions. They are comparable with the distinctions others make among givers and takers (Grant 2013b; Popova 2013).

The two previous images represent caring. Figure 2-3 illustrates indifference.

Figure 2-3. Indifference: Hungry child

We don't see who it is that is indifferent. It could be the child's family, or maybe the child's community, or the government of that child's country.

It might be that some of them are not indifferent, but simply do not have the capacity to relieve that child's plight. The fact that many parents in Haiti send their children to orphanages shows that some parents are unable to support their children (Brennan 2002). However, we do know that the global community, taken as a whole, has no such excuse. Global wealth in 2015 added up to US$250 trillion (Credit Suisse 2015). The world certainly has the capacity to feed hungry children. It cares, but not enough.

Globally, child mortality rates have been declining rapidly, but there are still more than six million children who die before their fifth birthdays each year. Most of these deaths can be attributed to global indifference rather than to any sort of direct abuse or exploitation.

Indifference is important. Many people suffer neglect, oppression, and violence and get no attention from the rest of the world. This is not only about international relations, how rich countries treat poor countries. Many countries are indifferent to the poor and powerless segments of their own populations. In Korea, suicides of people over 65 have increased sharply because their children no longer care for them adequately (Sang-Hun 2013).

Contrary to the image the United States presents of itself, almost half its people live in poverty (Edelman 2012). In India and the U.S., the poor get substantial subsidies from the government, but these programs seem to hold people in poverty rather than helping them climb out of it. One of the major impacts of the subsidies for the poor is that they ensure a steady supply of cheap labor, benefiting the rich.

In situations of economic exploitation, like that illustrated in Figure 2-4, much of the value produced by people's labor goes to benefit others, leaving the workers in marginal conditions. They are deprived not only in terms of income but also in terms of their dignity and their identity as distinct individual personalities.

Figure 2-4. Exploitation

Indifference and exploitation are important, but the focus here is on caring.

Sometimes caring takes on heroic proportions, as illustrated by war heroes and saints. We should also appreciate ordinary everyday caring. The caring of individuals for other individuals is often expressed in simple acts such as offering words of encouragement or just listening to others' concerns.

In a way, it is too bad that schools teach children about super-heroes like Mother Teresa, Gandhi, and Martin Luther King, placing them on sky-high pedestals. That elevation to heroic Nobel-prize status might make it appear that caring is out of reach for us ordinary flawed mortals, when in fact it is the stuff of everyday life, and within everyone's capabilities. The concern here is not about caring as a heroic rescue mission, but caring as a way of living together with others. We need to honor ordinary caring. It is the glue that holds civil society together.

Every decision says something about the decision-makers' priorities and their concern for others. When a report on global agriculture showed that production yield levels of some of the world's major food crops have been declining, one of the co-authors, Jonathan Foley, said,

> This finding is particularly troubling because it suggests that we have preferentially focused our crop improvement efforts on feeding animals and cars, as we have largely ignored investments in wheat and rice, crops that feed people and are the basis of food security in much of the world (Fisher 2012).

Every farmer and every manufacturer must make choices about what he or she will produce, just as every seller must choose what to sell, and every consumer must choose what to buy. In every action, each of us must make our own decisions regarding what we care about. Our decisions all add up to shape what it is that we as individuals, and also our social systems, care about.

QUANTIFYING CARING

There have been good efforts to design measures of well-being (nef 2016). Assessing the degree to which someone cares about others' well-being is a more difficult matter. The intensity of caring by individuals and groups can be estimated by using various indicators such as crime, charitable giving, and volunteerism. There have been some excellent studies of caring by individuals in laboratory settings (Singer and Ricard 2015), but few out in real-world communities. Few social scientists view communities as an important unit of analysis.

The degree to which a person cares about people and things can be assessed on the basis of the choices they would make in different circumstances. A person who likes vanilla ice cream more than chocolate chooses vanilla. The choice reveals the person's priorities. One way to estimate what individuals or agencies care about is to look at their budgets. If a community spends much more on football than on children's health, we have an important indication of what it cares about.

We can get clues not only from the allocation of money but also from the allocation of attention. If a country does not collect data on the nutrition status of its children, it is not likely to do anything about it (Vyawahare 2013). If the local media devote more space to stock market reports than to genocidal conflicts on the other side of the world, we have indications of what those media and their audiences care about.

Caring for another person's or group's well-being means making choices partly on the basis of how those choices affect that other person or group, and not solely on the choice's direct impact on oneself. Conceptually, the degree or intensity of caring could be estimated by assessing the extent to which an individual is willing to forego a benefit in order to deliver a given amount of benefit to another.

While caring does not always require sacrifice, the intensity of caring is indicated by the willingness to give up benefits to oneself in order to ensure benefits to others.

It is difficult to actually measure and study caring in quantitative terms, but the concept is important. Caring, acting to benefit another is essential to a well-functioning social order.

CARING IN COMMUNITIES

Caring is important to human well-being in every kind of community (Pilisuk and Parks 1986). It is especially important for people who are poor and therefore more dependent on the people around them. A poor person living among equally poor but caring people will have a much better quality of life than someone with the same income level living in an indifferent or exploitative community.

We tend to assume that people who have money are always happy, and those who don't have much money must be miserable. These assumptions certainly are not correct. People get a great deal of satisfaction out of their social interactions. There are many communities that do quite well despite their operating outside the dominant money exchange system.

Archbishop Desmond Tutu described the importance of group solidarity in this way:

> *Africans have a thing called ubuntu. It is about the essence of being human, it is part of the gift that Africa will give the world. It embraces hospitality, caring about others, being willing to go the extra mile for the sake of another. We believe that a person is a person through other persons, that my humanity is caught up, bound up, inextricably, with yours. When I dehumanize you, I inexorably dehumanize myself. The solitary human being is a contradiction in terms. Therefore you seek to work for the common good because your humanity comes into its own in community, in belonging* (Ubuntu Age 2012).

The theme was pursued at a conference held in Johannesburg in July 2013:

> *The theme of the event, "Caring Cities", is based on the concept of Ubuntu, an African ethic or humanist philosophy focusing on people's*

allegiances and relations. Caring Cities are cities that strive to offer a high quality of life, showing a sense of humanity and exchange, providing comfort and dignity for all citizens and deliver solutions that meet the needs of their citizens. (Joburg 2013)

Caring gestures can play important roles in defusing intense armed conflict situations. Sometimes they take peculiar twists, as in a counter-insurgency effort in India in October 1993:

In Srinagar, which is the capital of the State of Jammu & Kashmir, terrorists took over the most sacred mosque in the city. . . . The security forces laid siege to the mosque without actually attacking it in order to prevent damaging it. The attempt was to starve out the terrorists. Meanwhile someone sent in a Writ petition to the Jammu & Kashmir High Court stating that this was violative of the fundamental rights of the terrorists. The High Court decreed that the besieging forces could not inflict starvation on the terrorists and, therefore, they were directed to see that two full meals were sent into the mosque for its occupiers every day. It is only in India that we could have a situation in which the police and the army sent in two huge containers of food at noon and at sunset so that the terrorists could feed themselves. (Kent 2011a, 36)

The terrorists were seen as part of the community, deserving a measure of care.

Communities vary a great deal in terms of caring, indifference, and exploitation. Which are strong in the sense that people care about one another? Which are permeated by exploitative relationships? Which communities are marked by large-scale indifference to the poor or to other groups such as ethnic minorities or people with disabilities? Unfortunately, there are no well-developed indicators or data collections that offer insight on the quality of human relationships in communities.

Caring tends to be strongest in small face-to-face groups, but it can also be found in larger groups such as nations, spanning large stretches of both space and time. Caring can reach far. But of course no one expects that we could care for everyone on distant continents as much as we would care for our own families, and our own ethnic and religious groups. We rarely care for others who are distant geographically, chronologically, and socially in the same way we care for those who are close to us.

There are limits to caring, and counter-pressures that must be plainly recognized. In any intense, sustained conflict, one's acting together with people from the other side can be viewed as traitorous. This has been obvious in the Irish troubles, the Israeli-Palestinian-Arab conflicts, and many other situations. Often increasing one's connections with others can mean loss of connection with one's own people. This is why, in many hard conflicts, the two-state solution is best, with stable separation rather than common community. Instead of hoping for some sort of integration between the conflicting parties, it may be better for them to live apart, with arrangements that deter attacks or other de-stabilizing actions between them. Stable, respected boundaries are important means for protecting distinct groups.

Despite limits to our capacity for caring, we should care to some degree for all people who face extreme hardship and injustice, no matter how distant they might be. We should, but we don't.

A reasonable moral precept for dealing with distant communities is, at the very least, to do no harm. We cannot be expected to feed all those who are hungry in distant lands, but we should at least be sure to not take food or anything else away from them. In particular, we should not take the fruits of their labor to such an extent that they remain impoverished and unable to provide for themselves.

We should support the idea of social and economic safety nets, systems through which limits are placed on how far any person's well-being is allowed to decline. Local and global safety nets should be established, and over time they should be lifted to higher levels (Kent 2011a, 94-109).

ANIMALS AND PLANTS

The terms *caring, indifference,* and *exploitation* have been used here to describe relationships among people, but they also can be applied to other kinds of relationships. Many people act in a caring way to protect land, water, and animals in nature, even when those acts do not return any direct benefits to the actor. We recognize the exploitation of nature where people deplete and pollute their environment. Some farmers mine the soil and the water, making them unavailable for future generations, some fishers take too much, and some hikers wantonly destroy trees and trails, spoiling them for those who follow.

There are abundant examples of people being caring, indifferent, or exploitative toward animals. There are also many examples of animals treating people in those ways. Many animals display caring behavior to humans, often in homes, but also in other contexts. There are many examples of animals rescuing people in distress, inspiring many lists of such caring events (Crawford 2012; Frater 2010; Today Pets & Animals 2010).

There is clear evidence of caring between animals (de Waal 2009; Gandhi 2013), as illustrated below in Figure 2-5. Scenes of obvious caring are presented in photographs by Scientific American magazine (Wong 2012). A TV news story showed an extraordinary situation in which a lion, a tiger, and a bear lived and played together for years (ABC TV 2013).

Some animals show signs of grieving when their mates or friends die (King 2013). Surely this is a manifestation of caring. "Grieving profoundly ... is the price humans pay for caring profoundly" (Kluger 2013). That concept also applies to animals.

Figure 2-5: Caring in the world of animals

Caring among animals of the same species is commonplace, but caring across species is more interesting. We see it often in homes that have both cats and dogs. In some cases the two species merely co-exist, as separately as they can, but in others they interact actively and even affectionately.

Caring is more surprising when it happens between species of very different kinds, such as a cat and a bird or a rat and a cat (Chibudgielvr 2009a, 2009b). YouTube has many examples of inter-species connection, including some in the wild. These species are not "all one", but despite their diversity, they get along.

Some animals have uncanny skills in caring for humans. There are many stories of cats and dogs able to detect when a person is ill or in pain. Many work in hospitals and care homes as therapy animals.

Sometimes caring by animals for humans is not motivated by the animals' concern for the well-being of the humans, but is more of a business deal. The Eden Killer Whale Museum in Australia tells the story of a pack of killer whales that for decades helped to herd and kill baleen whales for the benefit of local human whalers. The trade-off was that the whalers would leave specific tasty parts of the baleen whales for the killer whales to feast on.

What about plants? There is a film that says forest plants live in communities of a sort, doing things to benefit one another (Public Broadcasting System 2013).

Michael Pollan says that when some types of closely related plants are placed in the same pot, they restrain their usual competitive behavior and instead share resources, and some trees in forest "communities" do similar things (Pollan 2013). Some researchers think corn plants show evidence of altruistic behavior (phys.org 2013).

Can plants really live in communities and demonstrate altruistic or caring behavior? Do the plants really care about one another, or is this merely a metaphor? What is the difference?

ARTIFACTS AND NATURE

We often draw lines between human-made artifacts and things in nature, but sometimes the boundary is not so clear. One basis for drawing the distinction is the idea that things in nature can have motivations, but inanimate things cannot. But maybe we make too sharp a distinction. We know cats and dogs have motivations because they can be observed doing things that are obviously goal-directed. However, for other life forms, the point is not so clear. Do bacteria have motivations? What about plants? Is it sensible to say that a bridge, a building, or a tool has a purpose of its own, independent of the users' purpose? Or is it better to just say these things are useful to us humans?

Cars can be much more than tools for getting from one place to another. Some people spend more money on dressing up their cars than on dressing their children. The uses of things often go beyond simple usefulness.

Is this idolatry? Is it simple anthropomorphism, attributing human qualities such as motivations to inanimate things? Is this just a matter of mislabeling things, or should ideas like "care" and "purpose" be understood as having broader meanings than we are used to giving them? Pantheistic religious perspectives often speak of objects having their own inner purposes.

We can care for nature and we can also care for artifacts such as bridges, buildings, canoes, and aquaculture ponds. We can be awestruck by beautiful cathedrals just as we can be awestruck by beautiful valleys (Keltner and Haidt 2003). Transcendent experiences, often marked by moments of awe, have something to do with caring. Both are about being in touch with something beyond oneself.

Clearly, people can care for nature. But does nature also take care of people in some active sense? Or do we simply try to adapt to a wholly indifferent, mindless nature?

These concerns can be framed in terms of the Gaia perspective, the view that "humanity constitutes a living system within the larger system of our Earth":

> *Take the living system most intimately familiar to all of us: the human body. We've long known that our bodies behave as a community of cells, which are organized into organs and organ systems. The central nervous system functions as the body's government, continually monitoring all its parts and functions, ever making intelligent decisions that serve the interest of the whole enterprise. Its economics are organized as an equitable system of production and distribution, with full employment of all cells and continual attention to their wellbeing. The immune 'defence' system protects its integrity and health against unfamiliar intruders. It can be thought of as a kind of global political economy with organs as bioregional units, their different tissues as communities, cells as families or clans, and the organelles within cells as individuals. Physiologically we can see that the needs and interests of individual cells, their organs and the whole body must be continually negotiated to achieve the body's dynamic equilibrium or healthy balance. (Sahtouris 1998)*

From this perspective it makes sense to speak about the "metabolism of cities" and the ways in which cities meet their needs (Deelstra 1987).

There has been a good deal of advocacy for recognition of legal rights of animals, as illustrated by the Nonhuman Rights Project (Nonhuman Rights Project 2016).

Some scholars have argued in favor of legal rights for natural objects. The underlying idea is that objects in nature, and maybe even artifacts, have intrinsic interests, apart from the usefulness they might have to humankind, and therefore the legal system should protect them (Stone 1974; Stone 1987). In Aotearoa (New Zealand), rivers are recognized as legal persons, having rights (Kennedy 2012).

As people care for the environment around them, that environment also cares for them, in a sense. This is not a matter of simple reciprocity, as in a deal to cooperate. It goes deeper than that, an unspoken symbiosis among elements in a shared system, organs in an organism. This dimension of caring is difficult to explore, involving concepts that may be alien or even alienating. It is one thing to suggest that the earth and the cosmos are alive in some sense, involved in some sort of dynamic organic flow, as in Gaia thinking. It is another huge step to suggest that there is something akin to consciousness in that environment. And still another step to feel and believe that this disembodied consciousness cares about the well-being of you and me.

There are interrelations among the different types and levels of caring. Caring for another person or an animal or a plant may be a spiritual act, at least for some people. Some people care for the corn or the taro they grow not simply for the nourishment it provides, but also for the contact it allows with another dimension of the world, one that is alive in its own way.

One manifestation of long-distance caring is international humanitarian assistance to people in need, coming from governments or from non-governmental organizations. The long historical arc of international humanitarian assistance is documented in Michael Barnett's *The Empire of Humanity*, ranging from the antislavery and missionary movements of the nineteenth century to disaster assistance and peace-building efforts in modern times (Barnett 2012). Frequently, the motives behind these operations have been mixed, as they often are in relation to caring acts at the personal level. International humanitarian assistance often falls far short of what is needed, but it is there, indicating at least a degree of caring (Kent 2013).

We tend to think of caring in terms of personal relations, with family and neighbors, and with our local environment. But caring is important for the larger whole, all of humanity, all of the world, and maybe all of the cosmos. These are transcendent forms of caring, going well beyond what we see and experience directly.

People say, "we are all one", but what does that mean? It cannot mean we are all the same because we are not. People are different from one another, and from cats and birds. We should take this "we are one" to mean that we are all important parts of a larger whole, just as the organs of the human body are parts of a larger whole. The parts don't always get along well with each other, but getting along certainly is a compelling recommendation. We *ought* to care for the whole, through space and also through time.

As parts of the larger whole, we can think of the organs of the human body as wanting to take care of their host. In much the same way, we can think of the earth and its environment as being, in some sense, motivated to take care of the human residents. Some scientists dismiss this sort of thinking as "promiscuous teleology" (Kelemen, Rottman, and Seston 2012; Kelemen 2012). People with a more spiritual orientation would dismiss those scientists.

Caring has been defined here as *acting to benefit others*. Many things in nature and many different kinds of artifacts provide benefits to people. Does this mean they care? Is the essence of caring in the action, or the thought behind the action, the underlying intention?

Does it matter whether the idea of motivation in animals or inanimate objects is viewed as a fact or a metaphor? Ideas don't have to be true to be useful, as we learn from imaginary numbers and Aesop's fables. Scriptures of various kinds survive not because their stories have proven true but because they have proven useful. Some people argue that nature does not really care about us (Lightman 2014), but it makes sense for us to treat nature as if it did.

Ideas of the sort discussed here might seem strange, but they have been spreading and growing in many quarters. They reached a high level when, "On April 22, 2013, in commemoration of International Mother Earth Day, the UN General Assembly hosted its third annual interactive dialogue on Harmony with Nature" (Zeleznik 2013). There is a United Nations website on the topic (Harmony with Nature 2016).

PAST AND FUTURE, AND SPIRIT

We commonly talk about caring for people and things as if they were here and now, but there are important space and time dimensions to caring. We tend to care more for people who are closer to us, physically, culturally, temporally, spiritually, and on other dimensions as well.

People care for things from the past and the future. Some give constant attention to their ancestors, and also to the godly precursors of their human ancestors, all honored in their creation stories. They also give attention to their descendants, and do a great deal for them. This occurs not only at the family level but also at the community and cultural level (Westra 2006).

The idea of "paying it forward" embodies the idea of caring projected into the future. Karma Kitchen illustrates it:

> *Imagine a restaurant where there are no prices on the menu and where the check reads $0.00 with only this footnote: "Your meal was a gift from someone who came before you. To keep the chain of gifts alive, we invite you to pay it forward for those dine after you. (Karma Kitchen 2016; Mehta 2013)*

People who are interested in establishing sustainable social systems care about the well-being of people and the planet well beyond their own lifetimes. Finance people are not the only ones who invest in the future and act to benefit their heirs. But when the benefits of an action in the present are spread among many people in the future, the motivation to act is likely to be much weaker than it would be when action benefits those who are close in term of geographical, temporal, and social distance (Walsh 2013).

Some people look far into the future and serve it. They also go beyond time to their spiritual worlds, in forms of transcendent caring. The pyramids of Egypt and the world's great cathedrals, synagogues, temples, and mosques were envisioned by architectural masters who knew they would never see the fruits of their work. Notre Dame in Paris, St. Peter's Basilica in Rome, Il Duomo in Milan, and Sagrada Familia in Barcelona are good examples. They demonstrate transcendent caring, beyond time.

In September 2016 Pope Francis issued a message for the World Day of Prayer for the Care of Creation, calling on us to Show Mercy to our Common Home (Pope Francis 2016). Here, again, the sometimes mundane concept of caring become transcendent.

Gaia thinking focuses on the earth as a source of care for people. Consider the more cosmic caring embodied in the much-quoted passage commonly attributed to Hafiz, a 14th century Sufi poet:

After a million years of shining
The sun never says to the earth –
'You owe me."
Imagine a love like this,
It lights up the Whole Sky

The attribution to Hafiz has been challenged (Azizi 2009). Nevertheless, whether viewed as fact or metaphor, the message is clear: Provide sunshine for others, without expecting anything in return.

EVOLUTIONARY BIOLOGY

The difference between animals and inanimate nature like plants and mountains seems obvious. Animals are sentient beings, but the others are not—we think. But microbiologists and deep sea explorers tell us that there are some life forms that straddle that boundary. The close dependence between animals and inanimate nature become apparent when we reflect on the observation, "all flesh is grass".

We sometimes attribute life-like characteristics to groups of animals. Herds and flocks and nations often seem to take on a life of their own, with a kind of group consciousness that is somehow above and beyond that of the individual members. There is something very real about groups that move in unison, even though the connections among their members are not visible. They have something of a transcendent, spiritual nature.

We often distinguish between humans and animals, but of course humans can be viewed simply as another type of animal. All show caring behavior. Humans have certain distinctive characteristics, but so does every other species. All have distinct roles in the global ecosystem. Only one thinks of itself as the primary one.

Evolutionary biologists help us overcome our self-indulgent preoccupation with our own species, encouraging us to see that we are parts of a much larger whole. All the elements that make up that whole depend on each other and, in various ways, care for each other. The environment is not something in a glass terrarium that we observe from the outside. We are inside that system, a part of it, with major impacts on it.

There are many unanswered questions. According to Jonathan Sacks, Charles Darwin "was puzzled by a phenomenon that seemed to contradict his most basic thesis, that natural selection should favor the ruthless":

> *Altruists, who risk their lives for others, should therefore usually die before passing on their genes to the next generation. Yet all societies value altruism, and something similar can be found among social animals, from chimpanzees to dolphins to leafcutter ants.*
>
> *Neuroscientists have shown how this works. We have mirror neurons that lead us to feel pain when we see others suffering. We are hard-wired for empathy. We are moral animals. (Sacks 2012)*

Sacks summarizes a core finding of evolutionary biology: "we survive as members of groups, and groups can exist only when individuals act not solely for their own advantage but for the sake of the group as a whole." The idea that we are purely, narrowly, and exclusively self-interested is wrong. We are not ruthless and lonely savages. We are self-interested, but we are also concerned with the well-being of others.

There is a problem, however. Some people suggest that concern for the group as a whole is about *universal* concern for others. The reality is that we tend to be deeply concerned only about selected others. We are not "all one," but members of many different groups. We are tribal by nature.

There are many different bases for tribal divisions, such as culture, ethnicity, and religion. Many tribes are in conflict with others, and often these conflicts turn violent. Caring for one's own group is often linked to hostility to others. People working within a corporation may feel strong solidarity as they struggle to overcome their competitors. Soldiers in the front lines put themselves in harm's way not to protect their individual selves, but to protect their brothers in arms and their tribes back home. Killing is claimed to be for a noble, selfless cause. Conflict with some can be driven by caring for others. The complexity can be appreciated when we contemplate acts of terrorism fueled by compassion for animals (Nelson 2016). Caring is not always all good all the time for all parties.

Strengthened bonds of caring within tribes can lead to aggressiveness and sharp conflicts between tribes. No one has yet found a good way to control that, especially in the anarchic world of international relations, where nationalism has been a potent force for both good and evil.

Some people suggest that major conflicts are due to tribal differences such as religious differences. But differences do not necessarily lead to conflict. Diverse tribes can happily co-exist so long as they don't invade each other's space, just as many different plants and animals can find a sensible equilibrium, living together harmoniously in their ecological space. Serious conflicts arise only when one tribe tries to impose its will on others.

Sustained conflicts between tribes often lead to the fracturing of bonds *within* the tribes, turning what might have been a simple manageable inter-tribal conflict into an intractable one. In the Middle East, for example, negotiations have become extremely difficult not only because of the incompatible positions of the Israelis and the Palestinians, but also because the divisions within each of those groups make constructive negotiations impossible. Jonathan Sacks thinks religion "remains the most powerful community builder the world has known," but he does not discuss the long history of intense conflict between religious groups, or the fact that many major religions are divided within themselves and have distinct progressive and fundamentalist branches. The Crusades continue on, even if many in the New World don't understand that.

Some evolutionary biologists see cooperation as a way in which some tribes can win in a deeply competitive world. Thus, Martin Nowak distinguishes five mechanisms of cooperation that can serve as strategies for winning in various types of struggle (Nowak and Highfield 2012; Nowak 2012).

The evolutionary biology approach may explain positive behavior in some circumstances, but how would it explain indifference or negative behavior? Why would a young man kill 20 school children in Newtown, Connecticut? Why would anyone set fire to a home in Webster, New York so that he could kill the fire fighters who try to protect the home and the people in it (Robbins and Kleinfeld 2012)? Why would people deliberately steer their cars to kill turtles crossing the road (Collins 2012)?

Why is there such widespread neglect and abuse of women, children, the elderly, animals, and nature? Attributing such actions to mental health problems does not explain them. Are these the behaviors of losers in a Darwinian struggle for survival?

Some evolutionary biologists look for explanations for caring behavior in some kind of indirect payoff, such as gaining advantages for one's tribe or gene pool over others. However, as pointed out earlier, caring people like Mother Teresa and Saint Damien were not concerned with gaining advantages over others. The goal of people's living well together with dignity and with respect for each other and their environment seems enough to explain why most people treat each other well most of the time.

Nowak's five mechanisms do not explain nice behavior in a non-competitive world. Evolutionary biologists who examine the benefits of cooperation in the context of struggles of various forms implicitly assume there must be losers. Must there be losers?

What Sacks viewed as an unsolved puzzle faced by Darwin actually was addressed by Darwin late in life. His masterwork, *The Origin of Species*, focused on pre-human evolution. However, in his much ignored later book, *The Descent of Man*, Darwin said:

> *We have now seen that actions are regarded by savages, and were probably so regarded by primeval man, as good or bad, solely as they obviously affect the welfare of the tribe,—not that of the species, nor that of an individual member of the tribe. This conclusion agrees well with the belief that the so-called moral sense is aboriginally derived from the social instincts, for both relate at first exclusively to the community. (Darwin 1936, 489)*

Darwin concluded that in human evolution, morality and conscience were more important than the idea of survival of the fittest (Loye 2007; Loye 2016; Johnson 2013; Tudge 2013b). As one observer summarized:

> *Darwin specifically denies that "the foundation of the most noble part of our nature" lies "in the base principle of selfishness." This flies in the face of the prevailing evolution paradigm, however, a form of Neo-Darwinism in which sociobiologists are vigorously pushing the idea that even altruism must be understood as motivated strictly by selfishness. (Lampman 2000)*

Some environmental biologists' analyses begin with Darwin and the need to survive. But then their discussion is somehow transformed into one about winning—whatever that might mean. Casting the struggle as being about winning means one assumes a world of perpetual conflict, one in which there must be losers. However, any stable ecology demonstrates the possibilities for long-term peaceful coexistence among individuals and groups of many different kinds. We do not have to live in a world of winners and losers.

Why do we often emphasize competition rather than cooperation? A friend told me that when he went to school, children were used to studying together. When it was classroom exam time, if they didn't know the answer to a question, they would ask the child sitting in the next row. Then they would be scolded by the teacher, and were told that learning from another child was cheating.

What is wrong with asking your neighbor when you are facing a puzzle? In a school based on deep caring, "Tests would be taken by groups helping one another get to correct answers, rather than separating children and ranking one higher or smarter than the other after the tests" (Laenui 2013). Smartness can be viewed as something that is held collectively.

Why have environmental biologists become so preoccupied with the idea of winning? The Social Transformation Faculty of Saybrook University asked, why is it that we "glorify winning as the undoing of an enemy, rather than an opportunity for life-saving reconciliation?" They articulated another option:

> *We support the development of a culture of transformative personal, organizational, and social change that fosters and celebrates the highest human qualities and practices, including empathy, altruism, peaceful conflict resolution, and restorative justice. (Schulman 2012)*

It is possible to organize things so everyone survives, but there is no way to organize so that everyone wins over others. It is possible for everyone to live well. We sometimes face a savage, competitive world, but not always, and not necessarily. There is enough for everyone's needs but not for everyone's greed.

Why assume there must always be competitors or that it is important to have one's own tribe outdo others? The idea of living well together might be enough to explain and motivate caring behavior.

INCREASING THE CARING

How can we go beyond simply describing the phenomenon of caring to instead increase the caring? It is not obvious. Preaching at people to treat one another nicely might help a bit. Having parents and teachers and other role models act nicely certainly has a positive influence. What else could be done to increase the caring in significant ways?

One promising approach has been developed in a private school in Hawai'i. Since 1991, its Psychosocial Education Department has operated alongside the more conventional math, English, history and science departments with the specific purpose of building empathy skills (Leong 2012). Using a broad variety of group activities such as ropes courses, along with more conventional classroom teaching, it appears to be successful.

There is a Center for Compassion and Altruism Research and Education at Stanford University:

> *CCARE investigates methods for cultivating compassion and promoting altruism within individuals and society through rigorous research, scientific collaborations, and academic conferences. In addition, CCARE provides a compassion cultivation program and teacher training as well as educational public events and programs (CCare 2016).*

Harvard University has a project called Making Caring Common that "helps educators, parents, and communities raise children who are caring, respectful, and responsible toward others and their communities. (Making Caring Common 2016).

One writer advocates "outrospection," in contrast to introspection, and suggests the creation of a museum of empathy that would help visitors come to a deeper understanding of other people's lives (Krzarnic 2012; Krzarnic 2013).

Drawing from the work of the late psychiatrist M. Scott Peck, the Foundation for Community Encouragement offers workshops designed to build the sense of community (Foundation for Community Encouragement 2016).

There is one major key to increasing the caring: *Relationships are strengthened when people work and play together.* This is an important element that should be designed into new communities and strengthened in existing communities. Local orchestras, sports teams, community projects, and business cooperatives, for example, all tend to strengthen bonds among their members. In well-designed communities, where people work and play together in many ways, people are likely to care for each other and for the local environment in which they are embedded.

Bonding among people who work and play together can take place in the pursuit of virtuous ends, such as producing good music, or questionable ones, such as those pursued by the Ku Klux Klan or street gangs.

At the outset, the goals of participants in these activities might be based on narrow self-interest. You might help a neighbor build his barn based on reciprocity, the anticipation that at some later time he might help you with a project of your own. However, over time, the practice of doing things based on instrumental caring is likely to evolve into increasingly empathetic caring. You might come to like your neighbors and want to do things for them, even things for which there are no likely returns on the investment.

In many cases, people band together to challenge a common enemy, whether a colonizer, an invading army, or a virus. Often that cooperation is based on each individual pursuing his or her individual interests. But over time, the collegiality involved in the effort is likely to transform the relationships into forms that go beyond simply using others, to showing some affection for them. We see this in the solidarity among soldiers or football players.

Being together and doing things together is likely to increase the caring. The thought is alarmingly simple. Apparently it is true for many different types of caring. As Kenneth Worthy explains,

> Only by being in sensuous, embodied contact with the rich, vibrant, complex realm of nature in landscapes and seascapes, with the air, soil, and water around us, can we begin to fully understand and experience nature's needs and thus be in a reciprocal and caring relationship with it. (Worthy 2013)

New communities can be designed or existing ones can be modified with a view to building the caring. Community design efforts often focus on the physical arrangements, but attention can also be given to the ways in which well-arranged communities can strengthen social relationships.

We can define *strong communities* as those in which people care about one another's well-being. They might be comprised of people who live close to one another and interact regularly. Employment opportunities, housing, and other amenities could be located in a defined contiguous space, thus allowing many of the residents to work close to where they live. Each community could have a management body, and rules determined through highly participatory processes. Such communities could produce much of their own food, and manage energy, waste disposal and many other concerns at the community level. Such communities would strive for sustainability, resilience, and self-reliance.

In such communities, people are likely to care for each other and for the local environment in which they are embedded. But there are no guarantees. The character of the community that emerges from the planning process would depend on the views and values of the individuals who do the planning. If the planning group advocates good nutrition, healthy people, healthy environments, and caring, the plans it puts out would reflect that input. Once implemented, this sort of community is likely to strengthen and transmit those values.

There has been a great deal of discussion of how businesses might pursue the "triple bottom line," giving attention not only to profits, but also to people and the planet (The Economist 2009). These "three Ps" could be pursued not only in the design of businesses but also in the design of strong communities. If a group of like-minded people brought in all their best ideas, drawing on everything they could learn about the best and the worst of both pre-modern and modern worlds, and had few obstacles in already-existing arrangements, they would have the potential for doing wonderful things together.

Tracing the motivation to benefit others shows us a common thread through all types of caring, ranging from the obvious forms that occur between parents and children to the more transcendent forms of caring for people and things beyond direct observation. Caring provides the basis for genuine connectedness, through many dimensions.

Empathetic caring takes connectedness beyond the merely mechanical and instrumental to something that is more deeply satisfying for us, individually and collectively. We need to work at understanding and increasing caring.

Chapter Three

Motivations for Producing Food

Food is abundant, as we can see in any supermarket. The world already produces more than enough food, but its distribution is skewed. "A billion starve because the wrong food is produced in the wrong places by the wrong means by the wrong people" (Tudge 2013b). Why?

HEALTH OR WEALTH?

Many people produce food mainly for subsistence and health. That has been the primary driver for most of human history. In modern industrialized food systems, however, food is produced primarily to increase the wealth of producers, not to improve the health of consumers. For producers, the money value has become far more important than the nutritive value.

This is well illustrated in the history of islands (Chirico and Farley 2015). In pre-contact Hawai'i food was abundant, and people were healthy. Taro and other foods were produced to meet people's needs. Then settlers came along and decided to produce rice for profit. There was a large-scale shift from taro to rice production in Hawai'i in the 1860s.

The rapid displacement of taro by rice led the local newspaper to ask, "where is our taro to come from?" The close link between production and consumption was broken, and the separation between farming for nutrition and farming for money became clear.

The people whose taro supply was threatened were not the people who benefited from rice exports. The Great Māhele enacted in 1848 allowed non-Hawaiians to own land, opening the way not only for rice but also for large sugar and pineapple plantations, cattle ranching, and more recently, seed research and production. The move to industrialized agriculture to serve distant customers was driven by interest in the wealth of the producers, not the health of consumers. While food systems used to be driven by people's nutritional needs, now the primary driver is the desire to make money.

One can eat only so much taro, so the impulse to produce it was naturally limited. As the motivation shifted into the desire to make money, there is no satiation. More is always thought to be better. Other considerations, such as limiting pollution and depletion of the environment, are overridden. This shift of the motivation for agriculture from producing community health to producing private wealth has been global in scope (Kaufman 2012; Lindgren 2013; Rosenthal 2013; Tudge 2013a, 2013b).

AGROECOLOGY

Agroecology evolved to meet the needs of people and the ecosystems in which they were embedded, in sustainable—almost timeless—systems (Anderson, Pimbert, and Kiss 2015; Human Rights Council 2010; Lappé 2016; IPES-Food 2016; Oakland Institute 2015; Pimbert 2009). Ancient Mayan cities and ancient Constantinople have much to teach us regarding urban food systems and the application of basic principles of agroecology (Stockholm Resilience Center 2013).

Some writers define agroecology as the practice of working with nature in farming systems, emphasizing organic modes of production. The practice is frequently described in terms of its technology, but a broader understanding would include people and their social organization as part of the ecology. It would view the community-based social organization of food production as more natural than exploitative industrial modes of organization. Agroecology emphasizes caring and cooperation, while industrial modes of food production are driven more by interest in private wealth than the well-being of the community and the environment in which it is embedded.

PRE-MODERN AGRICULTURE TODAY

Historically, local non-industrial food systems had tight links between farming and nutrition. These systems still function in much of the world where farming is not tied to modern markets:

> *Only 30% of the world's food supply is produced on industrial farms while half of the world's cultivated food is produced by peasants. More than 12% comes from hunting and gathering while more than 7% is produced in city gardens. . . .*
>
> *There are about 1.5 billion peasant farmers on 380 million farms; 800 million more urban gardens; and 410 million gathering the hidden harvest of our forests and savannas; 190 million in animal husbandry and well over 100 million peasant fishers. Many of our world's farmers are women. Better than anyone else, peasant farmers feed the hungry . . . (Courtens 2012, based on ETC Group 2009)*

Another group estimates, "Smallholders are responsible for over 90% of all investment in agriculture and for up to 80% of all the food produced and consumed in the world" (Transnational Institute 2015). The Food and Agriculture Organization of the United Nations provides comparable information through its Family Farming Knowledge Platform (FAO 2016b).

Pre-modern forms of agriculture are alive and doing well in many parts of the world, but they get little attention. Their effectiveness in providing good food supplies has been well documented (Inter Pares 2004; Kuhnlein, Erasmus, and Spigelski 2009). These time-tested modes of food production are losing ground.

DISTANCE FROM PRODUCERS TO CONSUMERS

Back when almost everyone was indigenous (Rasmussen 2013), farmers were responsive to the needs of local communities. However, as travel and trade have grown throughout the world, many food producers became disconnected from their local communities. With the encouragement of trade, farmers scan the horizon for the highest bidders for their services. Local needs are bypassed.

Producers and consumers are separated not only by distance but also by layers of wholesalers, processors, and investors who all have their own distinct interests in the food system. They ship the products to the most lucrative markets, as illustrated by the global fish trade and the fruit and vegetable trade. In the industrial food system, food is directed to people with money, anywhere in the world, not to neighbors.

DISTORTED VALUE CHAINS

Value chain analyses in the food industry rarely consider health impacts, whether on producers, consumers, or the environment. They also ignore the value of job creation and the associated distribution of the economic benefits of food production. In an extreme example of the industrialization of food production, the advocates of automated dairy farming boast that such systems will help take hard labor out of dairy (Williamson 2016). They will do this mainly by reducing the need for any sort of labor. That is not something to boast about.

The demand for food has grown much faster than can be explained by population growth. Many people now consume far more than they need for an active and healthy life. The global obesity epidemic provides ample evidence. Food processors' main interest is in adding economic value to the product, so the system delivers too much highly processed food.

Many modern agricultural investors are outsiders who take over distressed local farms or see farmland itself simply as another commodity to be traded for quick profits. There is continuity between the actions of settlers and colonists a few centuries ago and modern land grabbers (GRAIN 2012).

Industrialized food production tends to exploit workers, the environment, and customers. The maldevelopment of modern agriculture is well illustrated in Guatemala:

Guatemala has one of the world's highest rates of land concentration, where 3% of private landowners – a white elite – occupy 65% of the arable land. Small farms (those with fewer than four hectares) occupy only 11% of agricultural land.

Poor indigenous farmers scrape out a living through subsistence agriculture, often on the poorest soils, while wealthy plantation owners, or latifundistas, benefit from an agricultural system based on international exports such as coffee, sugar cane and African palm oil – and cheap, mostly indigenous labour (Tran 2013).

On both the producers' and the consumers' sides, the system benefits the rich far more than the poor, steadily widening the gap between them. The dominant economic system does not care much for people without money (Kent 1993). And, the evidence is clear, it does not care much for people who are hungry.

Advocates of large-scale modern agriculture often justify it by claiming economies of scale. Rather than efficiency in production, the key advantage of large farms is that they have one owner profiting from the work of machines and many poorly paid laborers. This is incentive enough for the owners. In modern agriculture, the selection of what to produce and how to produce it depends on the ability to expropriate others' labor (Guo 2016).

SOLUTIONS WITHOUT PROBLEMS

This helps us understand the global drive for genetically modified organisms in agriculture. Monsanto (Monsanto 2016) and Bill Gates (Malkan 2016) argue that GMOs are important to ensure food security, but that is doubtful. As Colin Tudge explains:

Overall, after 30 years of concerted endeavour, ultimately at our expense and with the neglect of matters far more pressing, no GMO food crop has ever solved a problem that really needs solving that could not have been solved by conventional means in the same time and at less cost.

> *The real point behind GMOs is to achieve corporate/big government control of all agriculture, the biggest by far of all human endeavours. And this agriculture will be geared not to general wellbeing but to the maximization of wealth. (Tudge 2013b)*

GMOs are not essential to food security because food is not difficult to produce. The problem is that it flows to people with money, not people with unmet needs but little money. GMO technology certainly is not the answer to the hunger problem.

The growth of modern agriculture is mainly about increasing concentration of control and wealth, not about increasing productivity, efficiency, or sustainability (Tudge 2013c). The priority given to wealth rather than health can be illustrated in many ways, such as the inattention to health impacts by the promoters of baby foods (Baby Milk Action 2016; Kent 2015d), and by the ways in which the promoters of protein-rich foods ignore the simple fact that many people get far more than they need (Starling 2016).

Many large farms are profitable because they operate in unsustainable ways, externalizing many of their social and environmental costs.

Many receive subsidies from government. Some subsidies are explicit, in the form of cash and some are hidden in the form of government services such as road-building, marketing assistance, and research.

The subsidies provide incentives for overproduction of certain commodities, leading to distortions in the human diet and in the economic system. The overproduction of corn in the United States is well-known (Pollan 2007), and the global overproduction of rice has become more visible (Lobello 2013). The excess food does not go to those who need it. It is more likely to be turned into a cheap ingredient for processed food, go to animal feed, or used to produce biofuel.

There is no good reason to subsidize wealthy large-scale farmers. If the public policy objective is to improve human nutrition, far more would be achieved by subsidizing well-managed small farms that produce basic fresh foods (Wiggins and Keats 2013) and by supporting breastfeeding (Gupta 2013).

People who think the hunger problem can be solved simply by increasing food production, described by their critics as "productivists", generally ignore the question of how people with little money are to access the new increments in food supplies. They focus on agriculture, which means they ignore the fact that in high income countries consumer expenditures for food go mainly to processors, not farmers. The technological improvements they advocate are likely to benefit some farmers while driving others off the land. Frances Moore Lappé calls on us to "see through the productivist fixation that inexorably concentrates power, generating scarcity for some, no matter how much we produce" (Lappé 2011b).

Many people devote a great deal of effort to developing new agriculture technologies, arguing that they will be needed to feed the world in the not too distant future. Some might be helpful, but as Tudge says about GMOs, we don't really need those new technologies.

Many novel methods of food production such as GMOs and high-rise pig farms are typical of the broader category, *solutions without problems*. These "solutions" are likely to benefit their producers more than anyone else.

LINKAGE BETWEEN AGRICULTURE AND NUTRITION

Several global agencies have been working to strengthen the linkage between agriculture and nutrition (Herforth 2012; Jones and Ejeta 2016; SCN 2013; World Bank 2007). The Global Forum on Food Security and Nutrition hosted an extensive discussion on "Making agriculture work for nutrition" (FAO 2012). This can be seen as a move to counteract one of the major deficiencies of the global industrialized food system, the fact that it responds mainly to money and not to needs.

The international agencies ask how agriculture might make a stronger contribution to nutrition without asking how the two became separated. Why do they focus on how agriculture *investments* can be more nutrition sensitive (FAO 2015a), implying that the first requirement is to make a money profit? Why insist that improving nutrition through agriculture should be based on the premise that "Food systems provide for all people's nutritional needs, while at the same time contributing to economic growth" (FAO 2015b)? Obviously, any enterprise must be economically viable if it is to be sustained, but that is different from saying all food enterprises must be major moneymakers.

The international agencies rarely discuss the role of the food marketers and processors who come between the primary producers and the ultimate consumers. They don't discuss the ways in which these power-holders in the food chain often marginalize the primary producers working in the fields. Most of the power is in these intermediaries, and in many countries these intermediaries receive the largest share of the money that is spent for food. The roles of the intermediaries in the system should be made more visible.

The dominant market-oriented food system does not provide for all people's nutritional needs. It is mainly the rich, not the poor, who benefit from economic growth. The economic benefits flow upward, and so does the food. The poor feed the rich (Kent 1982). To address the hunger problem, rather than focus on economic growth, it might make more sense to focus on sustainable ways for producing good food for the local community.

MARGINALIZATION OF FOOD WORKERS

In the gap-widening process described in the following chapter, food workers are among the most marginalized, impoverished and hungry as a result of low wages or low prices paid for what they produce. Some explanations of poverty focus on globalization and the roles of international financial institutions such as the World Bank and the International Monetary Fund, but the dynamics of marginalization are easier to see in local situations. A farmer might discover that she can purchase a new fertilizer that will increase yields by ten percent. Despite bright hopes, this does not translate into a ten percent improvement in her quality of life. Before long, landowners, seedsellers, and fertilizer sellers come along to claim a share of that gain. Before long, the farmer is pressed back close to her prior status. If she has low bargaining power compared to those around her, her increasing productivity is likely to benefit others more than herself. As a result of her powerlessness, she is perpetually marginalized.

The process of marginalization of the weak is illustrated by the way in which rice farmers in Vietnam failed to benefit from the sharp increase in global rice prices in 2008. Although the prices that Vietnam's farmers were offered for their rice increased, the prices of all their inputs, such as fertilizer and fuel, increased as well, so they were no better off (York 2008).

The marginalization of food workers is clear in both low income and high income countries. In India, the median annual wage for farmers is about US$290, including the value of the food they consume. The system of subsidies locks them into producing low value-crops (The Economist 2016).

In the United States, the median annual wage for food workers in New York State is around $20,000, lower than for all other categories of workers (New York State Department of Labor 2016). The income levels of many farm and restaurant workers are so low they remain food insecure and eligible for government food assistance programs (Sustainable Food Trust 2016). Many work under degrading conditions (Oxfam America 2016).

The income gap in California has been widening rapidly, due largely to the low incomes of farm workers in the state (Bohn and Danielson 2016). Referring to government programs such as the Supplemental Nutrition Assistance Program (SNAP) and the Special Supplemental Nutrition Program for Women, Infants and Children (WIC), the lead author said, "Policies that help bring low-income Californians more fully into the labor force and increase earnings hold promise for reducing inequality because families at all income levels rely most on their earnings from work." However, it is not obvious that programs that enable more poor people to work for low wages really is a good thing. Programs like SNAP and WIC make it possible for employers to hire people for less money than they otherwise would have to pay. Thus the subsidy to the poor also has the effect of subsidizing the rich. It may also do harm to the health and well-being of the children of poor families.

Ironically, millions of people who are unable to meet their food needs are food workers. Some agencies have responded by devising interventions such as bio-fortified crop varieties in the hope of improving these food producers' nutrition status (Fiorella, Chen, Milner, and Fernald 2016). In many cases it would be better to get these people out of food production altogether. The problem is not the nutritional quality of their products, but rather the fact that these workers' incomes are extremely low. As *The Economist* put it, "For Indian farmers to escape poverty, there need to be fewer of them" (The Economist 2016). Instead of helping them in farming, some should be helped to get out of farming. They need good alternatives.

Or, instead of trapping them with subsidies to produce low-value crops such as wheat and rice, they could be supported in switching to crops that are more valuable in both economic and nutritional terms, such as vegetables, pulses, and fruits (Jones and Ejeta 2016). People who are poor should be subsidized, but in ways that empower rather than trap them.

It is not obvious why food producers whose main interest is maximizing their wealth would be receptive to recommendations on how their enterprises might strengthen their impact on nutrition. If emphasizing nutrition does not lead to increased profits, the recommendations are not going to be very compelling for them (Kent 1986).

Many people express surprise and dismay at the amount of food that is wasted in industrial food systems. Once it is recognized that the primary objective of such systems is to maximize cash value, not nutritive value, that waste of food is no surprise at all. We need to come to a better understanding of why food producers do what they do, and get beyond naïve assumptions about the extent to which they are driven by the desire to achieve good health for all.

Chapter Four

Widening Gaps

As shown in Chapter Three, societies that center on market transactions and the building of wealth are very different from those that center on human relationships and the building of strong communities. In market systems, there is diminished care for the well-being of the people who are marginalized by it. This is closely linked to widening gaps between the rich and the poor.

EVIDENCE

The gap-widening is clearly demonstrated in the United States, where the middle class is being squeezed. Middle-income Americans are no longer in the majority (Fletcher 2015; Pew Research Center 2015; Woodiwiss 2013). Some observers seem surprised (Mason 2015), but steadily widening economic gaps between rich and poor were already signaled by Peter Edelman (Edelman 2012), Thomas Piketty (Piketty 2012), Joseph Stiglitz (Stiglitz 2013), presidential candidate Bernie Sanders, and many others. The video, *Requiem for the American Dream*, presents Noam Chomsky's analysis of the widening gap. The shrinking of the middle class in the United States has continued at least since 1971 (Ross 2016). In the gap widening process, increasingly large shares of the wealth that is produced go to people in the higher income categories, and increasingly large numbers of people move to the lower income categories.

Widening economic gaps have serious impacts not only on wealth but also other quality of life indicators, especially health. Angus Deaton, who won the Nobel Prize in economics in 2015, has shown, together with his wife, the decline in health of the United States middle class (Case and Deaton 2015; Deaton 2013; Kolata 2015).

It is not surprising that richer people have longer life expectancies than the poor. But the fact that the life expectancy gap between them is expanding should be a matter of serious concern. Just as wealth tends to flow toward the rich, gains in life expectancy also tend to favor the rich (Ehrenfreund 2015; Tavernise 2015).

This pattern shows up not only in the mortality rates of the elderly but also those of infants. The saving of lives of infants born prematurely has progressed much more rapidly in rich countries than in poor countries (Partnership for Maternal, Newborn & Child Health 2012).

In the United States and elsewhere, the general pattern is that those who start out with more wealth tend to gain wealth faster than those who start out with less. The annual increase in income of a corporation's senior officers is likely to be far greater than their janitors' annual salaries. Thus gaps steadily widen.

We see this pattern of divergence in the growth of nations. Those that start higher rise faster, while those that start lower rise more slowly. This is not obvious when people talk only about rates of growth. To illustrate, in the 1965-1984 period low-income economies had an average annual growth rate in their GNP per capita of 2.8 percent; lower middle-income economies grew at 3.1 percent; upper middle-income economies grew at 3.3 percent; and high-income industrial market economies grew at 2.4 percent (World Bank 1986, 180-181).

In these terms, it might appear that the growth rates were comparable, with the industrial economies growing at a slightly lower rate than the low-income nations. However, these figures are percentages of very different baseline levels of GNP per capita. We get a clearer picture if we compare the amounts, not the rates, of increase. In 1984, for example, low-income economies had average per capita income gains of $7.28, while industrial market economies had average per capita income gains of $274.32. Incomes per capita in industrial market economies were more than 37 times those in low-income nations.

In more recent times the per capita income in low-income countries grew at an average rate of 3.5 percent per year, while in high-income countries it grew at only 1.1 percent year (World Bank 2015b). But if we compare the amounts (not rates) of increase, we see that low-income countries had per capita incomes averaging $629 per year while in the high-income countries per capita incomes averaged $38,274. In the 1980s per capita incomes in high-income countries were about 37 times what they were in low-income countries, and in the 2010s per capita incomes in high-income countries were more than 60 times what they were in low-income countries.

Many people think of average per capita income as a primary indicator of national development. However, in terms of per capita economic growth *it is really the high-income countries that are the developing countries*. This might be among the unspoken reasons why the World Bank has stopped using the term "developing country" in its reports (Fernholz 2016).

Describing low-income countries as developing is misleading. OXFAM describes the pattern: "Far from trickling down, income and wealth are instead being sucked upwards at an alarming rate" (Oxfam 2016). Wealth bubbles up more than it trickles down.

EXPLANATIONS

The facts about the skewed distribution of wealth within nations and globally are clear. However, the explanations are not so clear. Why does this happen? The media carried many reports on the Pew Research Center's findings, but few ventured any explanation for the shrinking of the middle class.

Piketty says, "When the rate of return on capital exceeds the rate of growth of output and income . . . capitalism automatically generates arbitrary and unsustainable inequalities" (Piketty 2014). That has been the dominant pattern in the United States. Similarly, OXFAM says the economic system is rigged to work in the interests of the powerful through devices such as tax havens (OXFAM 2016, 4) That is true. But the economic gaps would widen over time even in the absence of such special arrangements.

While many people look to education as a means for enabling people to climb out of poverty, the education available to the poor, and the ascent it fuels, is meager when compared to the education and ascent it offers to the rich. The pattern is evident when the experience of students at Yale is compared with that of students at Norwalk Community College in Connecticut. Norwalk is one of the many local colleges that maintain food banks to help keep their students in school (Winne 2016).

Various factors contribute to the widening of economic gaps, but somehow many observers miss the most fundamental, straightforward, and comprehensive explanation.

The elementary transaction of the market system is the trade, the negotiated exchange. One's bargaining strength depends on the quality of one's alternatives. Some people (or companies, or nations) are stronger than others because they have better options.

Those who have greater bargaining strength tend to gain more out of each transaction than those who have less bargaining strength. Thus, over repeated transactions, stronger parties systematically enlarge their advantages over weaker parties. Traders do not move to an equilibrium at which the benefits are equally distributed, but instead move apart, with the gap between them steadily widening. Asymmetrical exchange feeds on itself, making the situation more and more asymmetrical. Whether we are speaking of people or nations, the incomes of the wealthy grow faster than those of the poor.

The market system is based on voluntary transactions in which both parties get some benefit. Any party that did not benefit could refuse the deal. Both parties benefit in the exchange process, *but unequally*. The rich (with greater bargaining power) get richer and the poor (with less bargaining power) get richer too, but much more slowly.

When the exchange process is accompanied by inflation, the real gains to both parties are diminished. The gains to the poorer, weaker party, being smaller, may as a result become negative. This is especially likely because inflation rates tend to be higher for poor nations than for rich nations. Thus with the combination of trade plus inflation it *is* likely that the rich get richer and the poor get poorer. The apparent gains from trade for the poor are likely to be wiped out by inflation.

There is no easier way to take away people's hard-earned money than by increasing prices across the board, which means decreasing the value of money. Alan Greenspan, former chair of the United States Federal Reserve System, described the process as "confiscation through inflation" (Greenspan 1966). Many people believe that inflation is deliberately manipulated to favor the rich (Griffin 2010).

UNEQUAL OPPORTUNITIES

The gap between rich and poor widens partly because transactions between them tend to be of greater benefit to the rich. However, the gap would widen even without those direct transactions. Those with larger amounts to invest always can get higher rates of return on their money. In a market system, the rich always have more and better alternatives, and thus enjoy faster economic growth than the poor. Wealthy people also have far better educational opportunities. They also have greater influence on the tax system, so they are able to promote tax policies in their favor (Scheiber and Cohen 2015). Power disparities lead to increasing wealth disparities.

TECHNOLOGICAL ADVANCES

There is a tendency to believe that innovations in technology benefit everyone, but most yield greater benefit to those who are already better off (World Bank 2016). Thus advances in technology generally contribute to widening the gap between rich and poor.

Those who start out with more tend to gain faster than those who start out with less. This holds not only not only with regard to wealth but also with other things of value, such as breastmilk. In the United States, infants from families with lower incomes are less likely to be breastfed, with serious long-term impacts on their health. One study concludes, "affluent mothers breastfeed, poorer mothers often rely on formula, and the cycle of inequality worsens" (Rodrigue and Reeves 2014).

Poverty is endlessly recreated. It is a product of an ongoing process, not a static condition or an original state of nature. If it were not, then surely, with all the anti-poverty programs that have been undertaken, it would have been eradicated by now. Those with low bargaining power are destined to remain marginalized because those with whom they interrelate have greater bargaining power.

Driven by the market system, the gap between the rich and the poor in the United States and elsewhere widens steadily, emptying out the middle class. That should not be surprising. There is nothing in economic theory or experience that would lead us to believe that a normally functioning market system would lead to a convergence in people's wealth levels. Corrections to this dominant tendency always come from outside the market system. Those corrections will be small if the powerful care little about the well-being of the powerless. Welfare programs are correctives, but often they are designed mainly to silence the poor, rather than provide them with a way out of their conditions. This is well illustrated in India, where many millions of people have pushed themselves into the Below Poverty Line category so they could get allocations of heavily subsidized grains.

We should expect steadily widening gaps between rich and poor in any market-based economy, local, national, or global. The middle class gets hollowed out. Somehow, this important characteristic of market-based social systems is not mentioned in the economics texts.

Chapter Five

Caring About Children

Many children live in wretched conditions. Some are physically abused in forced labor, armed conflict, and other bad situations. Many do not get the food, attention, health care, shelter, and other things they need. Many die.

CHILDREN'S MALNUTRITION

Malnutrition is a major factor contributing to the high mortality levels of infants and young children. Globally, the primary source of data on children's malnutrition is the United Nations Children's Fund, which works with the World Health Organization and the World Bank to provide an overview of levels and trends. They estimated that globally in 2014, of the 667 million children under five in the world:

- 159 million were stunted (under-height for age and gender).

- 50 million were wasted (under-weight for age and gender).

- 41 million were overweight.

(UNICEF-WHO-World Bank Group 2015; also see Roser 2016a)

The malnutrition that shows up as overweight has been increasing at an alarming rate in many populations, usually linked to increased consumption of processed foods. The tendency toward being overweight often begins in childhood (Bassi 2014; Cunningham, Kramer, and Venkat Narayan 2014). Infant formula, many people's first processed food, might be a significant factor leading to overweight in childhood and throughout the lifespan (Rose, Bodor, and Chilton 2006; California WIC Association 2006).

COMMERCIAL BABY FOOD

Many countries once categorized as poor are now described as "emerging economies" with a substantial middle class. These subgroups have money to spend, thus attracting sellers of many different kinds of goods, including foods for young children. There has been rapid growth of commercial, manufactured foods for young children, including not only infant formula, but also various follow-on (toddler) formulas, and semi-solid starter foods in packages promoted as more convenient than fresh foods (Baker et al. 2016).

Some of these products and their sellers are from inside the country and some from outside. Much of the manufacturing and marketing of these foods is based on joint-venture partnerships involving both insiders and outsiders. In many places, localized food systems are being overrun by outside interests. The accountability that once was based on direct contact between food producers and consumers is evolving into a global food system that is accountable to no one.

Much of the infant formula is promoted and supplied through international joint ventures. Cow's milk and other basic ingredients might be sourced in one place, and the formula might be manufactured somewhere else, with all of that controlled, more or less invisibly, by joint ventures headquartered elsewhere:

> *In just the five-years between 2008 and 2013 world total milk formula sales grew by 40.8% from 5.5 to 7.8kg per infant/child, a figure projected to increase to 10.8kg by 2018.*

> *We have described this recent surge in formula sales as indicative of a global 'infant and young child feeding transition' i.e. a shift from lower to higher formula diets at the population level. Although the idea of such a transition is not new, the rate and scale of change we describe is potentially unprecedented.* (Baker 2016; also see Kent 2015d)

The importance of breastfeeding, well documented for many decades (Williams 1939; Mata 1978), was confirmed again in a recent set of reports (The Lancet 2016; also see Stuebe 2009). There are serious national commitments to support breastfeeding (UNICEF 2013).

Consumption of infant formula is worse for infants' health than breastfeeding. While a good deal of attention has been given to concerns about the safety of using infant formula, little attention has been given to its nutritional adequacy (Kent 2011b, 2012a, 2014c; Minchin 2015a, 2015b). The inadequacy of formula is plainly evident from the fact that in any population, health outcomes for formula fed children are consistently worse than those for breastfed children. The United States Food and Drug Administration and other regulatory agencies don't seem to care much about whether infant formula is nutritionally adequate.

Measures have been taken to limit the inappropriate marketing of formula. In 1981 the World Health Assembly adopted the International Code of Marketing of Breast-milk Substitutes (World Health Organization 1981). The assembly has approved numerous resolutions in subsequent years to clarify and strengthen the Code. The International Baby Food Action Network, a nongovernmental organization, plays a leading role in monitoring adoption of the Code worldwide. IBFAN, together with the World Health Organization and the United Nations Children's Fund published a report in 2016 on the status of the Code's implementation (Leimbach 2016; World Health Organization, UNICEF, and IBFAN 2016). They voiced their concern that "Clever marketing should not be allowed to fudge the truth that there is no equal substitute for a mother's own milk" (World Health Organization 2016c; also see Rowland 2016).

In 2016 the World Health Organization issued a report on progress in carrying out its plan on maternal, infant and young child nutrition. Good progress has been made on many dimensions, but serious challenges remain. For example, despite the recommendation that infants should be exclusively breastfed for the first six months, complementary foods and breast-milk substitutes are being promoted as being suitable for infants under 6 months of age. Breast-milk substitutes are being indirectly promoted through association with commercial complementary foods. Inaccurate claims are being made that products will improve a child's health or intellectual performance. There is also concern that "complementary foods have been shown to displace the intake of breast-milk if the amounts provided represent a substantial proportion of energy requirements." (World Health Organization 2016e, para 33).

Following an all-too familiar pattern, efforts to regulate the marketing of processed foods for children were blocked by industry representatives and governments supporting them at a meeting of the World Health Assembly in June 2016 (Sterken 2016; Swinburn 2016; also see Kraak et al. 2016).

Manufacturers use claims about health benefits from using their products as a means for promoting sales. Many of the claims are vague and not supported by clear scientific evidence. The manufacturers do not present independent scientific studies to validate their claims. The challenges are well illustrated by the highly questionable health claims relating to fatty acids in infant formula and other baby foods (Kent 2014b).

Food manufacturers do not support sustained monitoring by independent agencies of the health impacts of the use of their products. In the case of infant formula, they do not do the studies they are supposed to do under the rules of the United States Food and Drug Administration. There are no publicly available data on the quantity of formula consumed, so it is not possible to assess its health impacts (Grayson 2016, 223-225). Food manufacturers show instrumental caring for the well-being of the consumers of their products, but there is little evidence of empathetic caring.

Questionable actions by food manufacturers are regularly documented by the Center for Science in the Public Interest and also by the Corporate Research Project. See, for example, the Project's *Nestlé: Corporate Rap Sheet* (Mattera 2013). The regular promotion of unhealthy food was criticized in *Scientific American* (Mustain 2013). The International Baby Food Action Network focuses on it specifically in relation to baby food.

The manufacturers of formula and other commercial baby foods do care about children's health, but that caring is limited. They definitely want to avoid risks of harm from their products such as those that can result from contamination. The care about the health of children in these cases is correlated with their fears that the alarm raised might lead to serious reductions in sales. Their caring is instrumental, not empathetic. This is evident in the ways in which they make questionable health claims and resist regulation designed to protect children's well-being.

In some situations, manufacturers exert so much influence, they turn governments into their partners. This is especially clear in the United States where the government, while advocating breastfeeding, provides about half the infant formula used in the country at no cost to the families (Edwards 2016; Grayson 2016, 204-31; Kent 2011b, 2016b). This raises questions about the quality of caring about children's well-being, not only on the part of the manufacturers but also on the part of the government. The government is supposed to regulate these companies, not serve as their agent.

HUMAN MILK BANKING AND SHARING

When infants are not breastfed by their own mothers, human milk can be supplied to them through milk banking or milk sharing.

Human milk *sharing* is based on direct contact between the primary providers of human milk and the infant's caretakers. For example, advertisements may be placed in newspapers or on the Internet to link the human milk providers directly to the infant's mother or other caretakers. Wet nursing is another form of sharing, with no banking stage between the provider and the final consumer of the human milk.

Banking involves collecting human milk at a central place, the milk bank, and then having infants' mothers or other caretakers obtain milk from the bank. Usually some processing is done at the bank such as pasteurization and quality testing. The women who provide human milk are likely to be screened through questionnaires and interviews.

Generally, there is greater regulation and quality control for banking than sharing, so the focus here is on human milk banking.

There is potential for extending the reach of human milk banking so that it serves infants who are not critically ill but would benefit from human milk from women other than their own mothers. Many would be infants who otherwise would be fed with breastmilk substitutes such as infant formula.

Feeding infants with human milk from milk banks is not as good for them as getting breastmilk from their own mothers, for several reasons:

1. The milk is not as fresh as that obtained through direct breast-feeding.

2. Banked milk might not be carefully matched for the age of the child.

3. Banked milk cannot change in response to immediate short-term needs such as those associated with infections.

4. Pasteurization, freezing, and other sorts of processing might lead to significant deterioration of the milk.

5. Banked milk is likely to be fed through bottles rather than through arrangements supporting direct skin-to-skin contact.

Despite such concerns, it is certain that banked human milk is better for infants than formula. It is important to promote and support direct breastfeeding by the mother, or at least the provision of breastmilk expressed by the mother. But when alternatives to the mother's own milk must be found, it is breastmilk from human milk banks that is second best, not infant formula.

The banking of human milk is expanding rapidly in both high and low income countries, but most banks serve only critically ill infants. Many other children who are not breastfed by their own mothers could benefit from having milk from other women made available to them.

There are concerns of various kinds, but they can be managed (Nice 2010; Path 2013). The potential benefits of making human milk more readily available should be pursued, to push back against the promoters of infant formula. Just as breastfeeding is a sign of a mother's care for her child, making human milk available for infants who cannot get their own mothers' milk would be a clear sign of the community's care for them.

RESPONSIBILITY AND CARE

While it is easy to describe the widespread malnutrition of children, explaining why it happens is a different matter. Unquestionably, local poverty an important factor, and there are local food shortages. But the global food system produces more than enough food for everyone, and it has the capacity to produce much more.

Food from the agriculture-based food system is important, but when considering infants and young children it is important to also give full attention to the importance of breastfeeding. There is extensive literature on this, but the widely accepted recommendations for optimum breastfeeding are not always followed. The situation is particularly problematic because of the inadequate research on the nutrition status of very young infants (Patwari, Kumar, and Beard 2015).

We are social beings, providing support to and drawing support from the people around us. At times we need help to take care of ourselves, especially at the beginning and the end of our lives. Sometimes we interact to gain something, and sometimes just for the pleasure of it.

The care *of* children refers to the work of parents or other caregivers in meeting their needs directly. Here, caring *about* children refers to what more distant agencies do—or fail to do—to ensure that children thrive. Caring *about* children should support those who care *for* them. In the United Nations Children's Fund framework, inadequate caring *for* children is one of the underlying causes of children's malnutrition, while inadequate caring *about* them is seen as one of the basic causes

(Jonsson 2015; Levitt, Pelletier, and Pell 2008.) A great deal has been written about the care *of* children, some to provide guidance, and some to provide an overview of its importance (Engle, Menon, and Haddad 1997). There have not been any overview studies of caring *about* children.

Children are in training for independence. They need to have others take care of them. The first line of responsibility is with the parents, of course, but others have a role as well. The key issue is not whose fault is it that children suffer so much (who caused the suffering?), but who should take action to remedy the situation. Many different social agencies have roles to play: families, communities, churches, nongovernmental organizations, businesses, and local and national governments. It is important to be clear about the interrelationships of their responsibilities for children.

Most children have two vigorous advocates even before they are born. Most parents devote enormous resources to serving their children's interests. These are not sacrifices. The best parents do not support their children out of a sense of obligation or as investments. Rather, they support their children as extensions of themselves, as part of their wholeness.

In many cases, however, that bond is broken or is never created. Fathers disappear. Mothers disappear as well. In some cities hundreds of children are abandoned each month in the hospitals in which they are born. Bands of children live in the streets by their wits, preyed upon by others. Frequently children end up alone as a result of poverty, disease, warfare or other sorts of crises. Many children are abandoned because they are physically or mentally handicapped. Some parents become so disabled by drugs or alcohol or disease that they cannot care for their children.

In many cases the failures are not the parents' own fault, but result from the fact that others have failed to meet their responsibility toward the parents. There are cases in which parents are willing to work hard and do whatever needs to be done to care for their children, but cannot find the kind of employment they need to raise their children properly.

Many children who cannot be cared for by their biological parents are looked after by others. In many places orphanages have disappeared, replaced by systems of formal or informal foster care. In some cultures, children belong not only to their biological parents but also to the community as a whole. The responsibility and the joy of raising children are widely shared.

In many places that option is no longer available because of the collapse of the idea and the practice of community. People may live in nice neighborhoods in well-ordered societies, but the sense of community—of love and responsibility and commitment to one another—may be missing. In such cases the remaining hope of the abandoned child is the government, the modern substitute for community. People look to government to provide human services that the local community no longer provides.

As children mature the first priority is to help them become responsible for themselves. So long as they are not mature, however, children ought to get their nurturance from their parents. Failing that, they ought to get it from their relatives. Failing that, they ought to get it from their local communities. Failing that, they ought to get it from the local governments. Failing that, it should come from their national governments. Failing that, they ought to get it from the international community.

The responsibility hierarchy can be portrayed as a set of nested circles, as suggested in Figure 5-1 below. The child in the center of the nest is surrounded, supported, and nurtured by family, community, government, and ultimately, international organizations.

There are exceptions to this nesting rings pattern. In some situations it makes sense for central governments to provide services to children or other needy people directly, bypassing local government and local nongovernmental agencies. Some programs, such as immunization, cannot be completely managed locally. Some services maybe distributed more equitably if they are funded out of the central treasury, rather than by the local community. In the United States, the responsibility for educating children is left to the states. Given the uneven wealth levels of the states, this has had the unfortunate effect of ensuring huge variations in the quality of education. This localization perpetuates poverty.

Figure 5-1. Rings of responsibility.

Nevertheless, with some important exceptions, the general pattern is that we expect problems to be handled locally, and reach out to more distant agents only when local remedies are inadequate. Subsidiarity serves as a guiding principle. To the extent feasible, issues should be handled locally, and raised to higher levels only when they cannot be managed locally.

In cases of failure, distant agents should not simply substitute for those closer to the child. Instead, those who are more distant should try to work with and strengthen those who are closer, in order to help them become more capable of fulfilling their responsibilities toward children. Agencies in the outer rings should help to overcome, not punish, failures in the inner rings. They should try to respond to failures in empowering, positive ways. To the extent possible, local communities should not take children away from inadequate parents but instead help them in their parenting role. Higher level agencies should not displace local agencies, but instead should support local agencies in their work with children. The international community should help national governments in their work with children.

In ordinary circumstances, government's responsibilities with regard to children should be limited. The family should provide daily care and feeding. However, for children in extreme situations who are abused or who suffer from extremely poor health or serious malnutrition, governments have a role to play. If there has been a failure in the inner rings of responsibility and no one else takes care of the problem, government should step in.

Empowerment—or development—means increasing one's capacity to analyze and act on one's own problems. Thus, empowerment is about increasing autonomy and decreasing one's dependence on others. The concept applies to societies as well as to individuals. It certainly applies to children.

There are similar rings of responsibility for others who cannot care for themselves, such as victims of disasters, the physically disabled, and mentally ill. These responsibilities need to be clarified so that the care of those who are unable to care for themselves is not left to chance. This conceptual framework can be used in relation to all individuals who need protection and support, and not only children.

Fulfilling children's nutrition needs is primarily the responsibility of their parents, their extended families, and the communities that surround them. There is a need to strengthen the chain of caring so that families and their communities become more capable of caring for their children. The hierarchy represented in the rings of responsibility presented is idealized, more wishful than descriptive.

OPTIONAL PROTOCOL ON CHILDREN'S NUTRITION

As discussed in Chapter Six, the Food and Agriculture Organization of the United Nations, the United Nations Children's Fund, the World Health Organization, and the World Bank do a great deal on food issues. However, the Committee on World Food Security, the United Nations' apex organization on food security, gives little attention to children, despite their being the most nutritionally vulnerable segment of the population. The Committee's meetings and the annual reports on *The State of Food Insecurity in the World* rarely discuss the distinctive needs of children. The Committee should take a leading role in bringing these concerns to the world's attention.

The Committee, working with national governments and international agencies, could launch an initiative to bring children's nutrition issues into sharper focus and clarify the division of responsibilities. This could be done through the global human rights system. With the support of the United Nations, the countries of the world could negotiate a new Optional Protocol on Children's Nutrition, to be linked to the United Nations Convention on the Rights of the Child (Kent 2016).

That Convention already has two Optional Protocols associated with it, one on the involvement of children in armed conflict, and another on the sale of children, child prostitution, and child pornography. Their forms could be used as a basis for drafting a new Optional Protocol on Children's Nutrition.

Working under the auspices of the United Nations General Assembly, the nations of the world could negotiate a draft Protocol. The drafters could draw from the many sources of sound principles relating to children's nutrition such as the International Code of Marketing of Breast-milk Substitutes, the Global Strategy for Infant and Young Child Feeding, the Innocenti Declaration, and the Baby-Friendly Hospital Initiative. Ideas from many other agencies and documents, now scattered, could be pulled together.

The Protocol could be limited to broad statements of principles and linked to a detailed General Comment on Children's Nutrition. It could be modeled on General Comment 12 on The Right to Adequate Food. General Comments provide authoritative interpretations of the law in formal human rights treaties

Most discussion of human rights focuses on the obligations of states to their own people. However, if human rights are global and not merely national, it must be recognized that all countries have obligations to people outside their own countries. Those obligations need to be clarified. The drafters of the Protocol could open a global discussion about the moral and legal obligations of the global community, taken as a whole, to all children everywhere.

When a draft for the Optional Protocol was ready, the General Assembly of the United Nations would vote on it. If a majority agreed, it would be adopted by the General Assembly. The executive branches of the national governments of the world would then be invited to sign the Protocol, and then have their national legislatures or other appropriate bodies ratify it, in the normal procedure used to formalize a nation's agreement to international treaties.

Ratification would indicate the nation's acceptance of the Protocol and its commitment to conform its national laws to the Protocol. Following ratification, its broad principles would be given concrete form through the adoption of appropriate national laws. The ratification would signify the nation's willingness to be held accountable with regard to the principles stated, and the new national laws would be the means by which its leaders follow through on the nation's commitment.

An Optional Protocol would not generate quick solutions to the many major issues relating to the nutrition of infants and young children. But the negotiation process would draw attention and launch much-needed discussions of the issues. The process could lead to substantial positive impacts even before the document is finalized.

If the challenge is taken seriously, the new Optional Protocol on Children's Nutrition, linked to the Convention on the Rights of the

Child, would help to establish coherent guidelines and regulations for ensuring that infants and young children everywhere are well nourished. The main thing higher level agencies need to do is provide support for individuals and agencies closer to children so they can do a better job of caring for them.

Chapter Six

Global and National Actors

GLOBAL ACTORS

The most prominent intergovernmental organizations concerned with food and nutrition are the Food and Agriculture Organization of the United Nations, the World Food Programme, the International Fund for Agricultural Development, the World Health Organization, and the United Nations Children's Fund. The United Nations Educational, Scientific, and Cultural Organization is also involved. They are governed by boards composed of UN member states. The World Trade Organization and the Codex Alimentarius Commission also play important roles, as do many transnational corporations.

Responsibility for coordinating food and nutrition activities among these and other global agencies rests with the United Nations System Standing Committee on Nutrition. Representatives of bilateral donor agencies such as the Swedish International Development Agency and the United States Agency for International Development also participate in activities of the Standing Committee on Nutrition. A number of global nongovernmental organizations are heavily involved in nutrition issues, including the Global Alliance for Improved Nutrition, known as GAIN, and The Hunger Project.

There is a Committee on World Food Security, funded by the Food and Agriculture Organization of the United Nations, the International Fund for Agricultural Development, and the World Food Programme. It is "the foremost inclusive international and intergovernmental platform for all stakeholders to work together to ensure food security and nutrition for all" (Committee on World

Food Security 2016). Its mandate overlaps that of the United Nations System Standing Committee on Nutrition.

Global level data on food and nutrition issues are provided by the Food and Agriculture Organization of the United Nations, the United Nations Children's Fund, and the World Health Organization. Several global nongovernmental organizations present data as well. National level data are reported by many national governments.

The United States Department of Agriculture also makes estimates of food security in selected countries, based on methods that differ from those of the United Nations agencies (United States Department of Agriculture 2016).

There is a Famine Early Warning Systems Network (FEWS NET 2016) created by the United States Agency for International Development to provide early warning and analysis on acute food insecurity.

The DuPont chemical company sponsors a Global Food Security Index designed by The Economist Intelligence Unit (The Economist Intelligence Unit 2016). The initiative supports the interests of DuPont and others global commercial interests in "a wave of consolidation in the agricultural seeds and chemicals industry" (The Economist 2016b). There is an emerging alliance of big food producers that hope to profit from widespread concerns about food security. Syngenta and its Good Growth Plan is part of it (Syngenta 2016) as is Monsanto (Monsanto 2016). Most corporations' concern about food security extends only to the reach of their business interests (Martin-Prével and Moussea 2016). The advocates of industrialized agriculture and globalization of the food system through free trade tend to give little attention to these systems' impacts on the well-being of the poor (Lincicome 2016).

MODEST ASPIRATIONS

The Food and Agriculture Organization of the United Nations is commonly thought to be the lead actor for ending hunger in the world, but its constitution does not establish this mandate. Instead, its constitution sets smaller goals, such as "the improvement of education and administration relating to nutrition, food and agriculture, and the spread of public knowledge of nutritional and agricultural science and practice."

The mission statement of the World Food Programme says, "The policies governing the use of World Food Programme food aid must be oriented towards the objective of eradicating hunger and poverty." Of course that approach alone is not enough. International food aid is usually provided in sudden-onset acute situations, providing temporary relief, but in many places hunger is a chronic condition. Food aid in emergencies is not going to solve the hunger problem.

The global agencies have modest aspirations in dealing with the global hunger problem. In the Millennium Declaration of 2000, 191 member states of the United Nations committed themselves to the first Millennium Development Goal, "to halve, by the year 2015, the proportion of the world's people whose income is less than one dollar a day and the proportion of people who suffer from hunger." The Millennium Development Project then reduced the broad aim of halving the proportion of people who suffer from hunger to the aim of reducing one particular form of malnutrition by half. Even so, the "strategy" that was finally offered could not have been expected to actually achieve that relatively modest goal (United Nations Millennium Project 2005).

The annual assessments of *The State of Food Insecurity in the World* prepared by the Food and Agriculture Organization, the International Fund for Agricultural Development, and the World Food Programme show that progress toward ending hunger has been slow (FAO, IFAD, and WFP 2015). Progress has actually been much slower than portrayed, as shown by Thomas Pogge (2004, 2016) and also by Jason Hickel:

> *The final report on the Millennium Development Goals (MDGs) concludes . . . that global poverty has been cut in half, and global hunger nearly in half, since 1990. . . . In reality, around four bil-*

lion people remain in poverty today, and around two billion remain hungry—more than ever before in history. *(Hickel 2016; also see Food First 2016)*

At the World Summit on Food Security of November 2009 and many other occasions, rich countries refused to accept a clear and firm hunger reduction target (Blas 2009). No actors can be called to account, individually or collectively, for failure to end hunger because none have made a serious commitment to achieve that goal.

There are examples of strong aspirations such as the World Health Assembly's goal of achieving a forty percent reduction in the number of stunted children by 2025, but the funding called for is not showing up (Generation Nutrition 2016). This is partly due to the funding agencies not caring enough about the problem and partly due to their doubts about the workability of the plan for achieving the goal (World Health Organization 2016d). There is also hesitancy because of concerns about possible misallocation of funds or outright corruption (GRAIN 2014; Kaufman 2009).

The United States Agency for International Development has a Food and Nutrition Technical Assistance Program called FANTA. It identifies a number of "essential intervention programs" and uses an activity-based costing methodology to estimate how much money would be required to implement those interventions. In Guatemala, for example, in 2013 it found that the government invested only about a third of what would have been required in that year. The agency is pleased that . . .

> *The results from the exercise are helping to raise awareness among key government stakeholders of the consequences of malnutrition for Guatemala and the need to increase investment in nutrition. (US-AID 2016a).*

Who in Guatemala is unaware of the importance of malnutrition in that country? There is no reason to believe that malnutrition there has persisted because government officials were unaware of it. It is difficult to find any basis for confidence that the FANTA approach will work.

Many activities funded under Feed the Future and related programs of the United States Agency for International Development seem to be devoted more to the promotion of industrialized agriculture than to the alleviation of malnutrition (USAID 2016b). They are not very empowering for those who are supposed to benefit.

Since 2012 global agencies have been coordinating their responses to all forms of malnutrition through the Scaling Up Nutrition program (Scaling Up Nutrition 2016). Its third *Global Nutrition Report* steers the program (International Food Policy Research Institute 2016). The document provides a comprehensive review of the current state of nutrition in the world and calls for specific commitments to end malnutrition by 2030.

The approach might seem comprehensive and ambitious, but it has important weaknesses. It has an overbearing top down-orientation. It is possible to have strong leadership from the top and exercise that leadership in a way that draws out the voices and the engagement of the people who are malnourished and who are supposed to benefit from all this effort. However, the *Global Nutrition Report* says little about what people who are malnourished might do for themselves. The needy are called on to be actors, but they are not involved in writing the script. Their voices are not heard in the document. The dominant theme is that the experts know what actions to take.

Many of the poorest countries do not participate in the Scaling Up Nutrition Program, which means it will not benefit malnourished people in those countries. Thus, from the outset, the program gave up on the idea of ending extreme malnutrition worldwide.

The right to food is mentioned only in passing, with no discussion of its potential or its limitations as a tool for helping to end malnutrition.

The preparation of the *Global Nutrition Report* involved not only the Independent Expert Group, but also a large number of "stakeholders," experts of various kinds from both high and low income countries. Most of them had no power to make commitments on behalf of their countries or their agencies.

Table 1.1, entitled Building a Global Commitment to Nutrition, lists the events since 2011 at which commitments have been made to address problems of malnutrition. Many other events could be added to the list (Kent 2011a, 169). Is the commitment really *building*?

The document ends with calls to action that are vague and not likely to be helpful to national or sub-national governments, or local communities. There is no plan that would lead one to expect achievement of specific goals related to the ending of malnutrition.

In April 2016 the United Nations General Assembly adopted a resolution proclaiming a UN Decade of Action on Nutrition from 2016 to 2025. It called for intensified action to end hunger and eradicate malnutrition worldwide, and ensure universal access to healthier and more sustainable diets–for all people, whoever they are and wherever they live. The language inspires, but the historical record raises doubts.

LONG DISTANCE CARING

The document's first "call to action" is that the world should "make the political choice to end all forms of malnutrition." Is that really plausible? Many global conferences have made similar calls, with little effect. Hope is not a plan.

The failure to make serious plans for the achievement of major nutrition goals says a great deal about the pinched quality of caring about malnutrition by the international agencies. However, this must be viewed with understanding. The intergovernmental organizations are servants of the member countries that set their agendas and fund their activities. The quality of caring by the global agencies reflects the quality of caring among the national governments and the funding agencies that direct them (Pogge 2016, 22).

There has never been a serious plan for ending global hunger. A serious plan is defined here as one for which adequate resources have been allocated and an effective management system has been set up in a way that gives observers confidence that the goal will be achieved. The problem is not that the global plans have failed, but rather that there has never been a serious plan for ending hunger in the world (Kent 2011a, 170-172). The absence of serious plans to end global hunger tells us that the motivation, the caring, is not as strong as it needs to be if it is actually to be ended.

The disinterest in hunger abroad became clear to me when a book I edited on *Global Obligations for the Right to Food* was published in 2008 (Kent 2008b). It was a time of global food crisis based on the rapid increase in prices of major food commodities. I thought the crisis would lead to great interest in the book, but there was zero interest. Apparently no one wanted to consider the idea that the global community, taken as a whole, should have clear legal obligations to help out people who were hungry. Others have made similar calls for clarification of obligations of countries to needy people everywhere, as distinguished from their obligations to their own people (cf. Felice 2013; Hoffman 1981), but that conversation has never gotten a serious start. Caring does not reach very far.

National governments give low priority to hunger among people outside their jurisdictions. We also see serious limitations in their caring for their own people. In the poorest countries the national governments say they are concerned about their people's hunger, and they invite outsiders in to provide food aid and other services. However, many of the governments in poor countries could do much more on their own, if they put higher priority on the issue. They may not care as much as they claim they do when asking for foreign aid.

Some countries are rich but have many poor people. The United States, Brazil, and India are major examples. People think of the United States as a rich country, but about a fifth of its children live in poverty. Half the infants in the United States get free infant formula from a government program that provides supplementary nutrition for people with low incomes. This is done in a way that benefits the formula manufacturers more than the infants (Kent 2006b, 2011b).

Inequality within these countries is very great, on many dimensions, especially with regard to children and the opportunities they have (Putnam 2016). Greed and corruption overwhelm caring, increasing the gap between rich and poor. As shown in Chapter Three, the dominant economic system in these countries ensures that gaps steadily widen.

India has been among the leading exporters of rice, wheat, and beef (actually buffalo) while at the same time a large part of its population suffers from serious malnutrition. Its Integrated Child Development Service is responsible for the well-being of children under six years of age. Despite the program's massive budget, India still has more malnourished children than any other country in the world. As a result of corruption, there is huge leakage of resources from the program. It would not be difficult to make it work more effectively, if the motivation was there (Kent 2012b).

The agencies in the United Nations system do good things to improve nutrition situations (Gillespie et al. 2016), but in many cases there is little lasting impact. One study found that a package of ten nutrition-specific interventions covering 90 percent of the countries with high levels of child stunting would be expected to avert only one-fifth of the stunting in those countries. The annual cost of doing this was estimated at $9.6 billion per year (Bhutta 2013).

Projects tend to focus on narrow goals such as improvements in the intake of particular nutrients in the short term. They look for readily measurable outcomes.

Some projects are said to be community-based (FAO 2005b; World Health Organization 2003), but often that refers only to their geographic location. Typically, the motivations, methods and resources come in from the outside, with the interveners.

Some interventions use manipulative technical procedures in which outsiders show little respect for local people's perspectives and dignity (ACDI-VOCA 2016). They say things like, "Under the AGP-AMDe project male farmers were targeted with appropriate analogy nutrition messages through a cascade training in primary cooperatives." This is not the language of small-scale Ethiopian farmers.

The manipulative approach to improving nutrition is deployed by GAIN when "they construct nutrition-relevant messages based on user 'personas' that can help to identify personal motivations, attitudes, desires, and other drivers of behavior change. These personas are then linked to text messaging campaigns that are validated by country stakeholders" (GAIN 2016).

Caring from the top levels is present, but thin. To some critics, it is little more than a matter of theater, based on the need to ensure proper appearances:

> *Citizens care about severe poverty and hunger, both at home and abroad. Therefore governments and politicians in power have an interest in expressing support for the struggle against poverty and hunger... . And through this joint performance governments also divert attention away from the structural causes for the persistence of severe poverty... . we citizens are led to overlook the very real fact that 2015 also saw the richest 1 % of humanity expand its share of global private wealth to over half (50.4 %). The poorer half of humanity, meanwhile, was squeezed down to a mere 0.6 % of global private wealth, as much as is owned by the world's richest 62 billionaires. (Pogge 2016, 21-22)*

The long-term pattern of responses from both international agencies and national governments tells us that on the whole they do not care much about hunger. We have to look at what they do, not what they say.

ALTERNATIVES TO INTERVENTIONISM

We do not need the manipulative or theatrical sorts of caring about hunger. Food and nutrition projects should focus on working together with local people in conversations about their local situation in which all involved come to a better understanding of the local situation (Kent 1988). This calls to mind the statement attributed to Australian

Aboriginal Elder Lilla Watson: "If you've come here to help me, you're wasting your time. But if you've come because your liberation is bound up with mine, then let us work together (Watson 2016)."

As discussed in Chapter Five in relation to the rings of responsibility, it is important to recognize that caring generally diminishes in intensity over great distances, whether that distance is geographical, social, religious, or something else. We cannot expect United Nations officials and experts in Rome or Geneva or New York are going to care as much about malnourished neighbors as those who live around those people. The strategies adopted by the high level agencies should take that reality into account.

This means getting beyond treating malnourished people as silent statistics, and instead getting a better understanding and providing better support for the local people so *they* become better able to address their issues.

There is an effort by the Food and Agriculture Organization to get beyond cold statistics to grasp how local people actually experience malnutrition, called "Voices of the Hungry" (FAO 2016c, 2016d). However, the purpose of the program is to develop a better understanding so that the interveners can make better decisions, not to empower the local people themselves. This is evident from "the theoretical background and detailed description of the statistical methodology used to develop the Food Insecurity Experience Scale (FIES), a new metric for food access at the household and individual levels." Having outsiders structure the conversation through questionnaires designed by the outsiders is demeaning to local people, signaling that they are not equal partners in the conversation. I have made similar comments elsewhere about the FAO's Food Insecurity and Vulnerability Information Management System, known as FIVIMS (Kent 2014a).

Insiders and outsiders should come to a good *shared* understanding of the local situation. They should work toward a more even relationship of learning from each other, as Lilla Watson might have imagined it. The local people are not there merely to "provide up-to-date information about food insecurity that is policy relevant" in a form that works for the outsiders but not for themselves.

Just as health does not depend only on medicine, good nutrition status does not depend only on issue-specific interventions. Nutrition status depends on the quality of the relationships among the people within the community and also on their relationships with others in and from other communities. Those relationships can be strengthened. Supporting and guiding local people in making their own assessment of their own local food situation in their own terms could be highly empowering for them, and help to strengthen the relationships among the local people.

Local communities can be the site of conventional nutrition interventions by outsiders, but in strong caring communities there will be much less need for such interventions.

Chapter Seven

Designing Caring Communities

We need to find a way to imagine, design, and implement a post-modern world that draws on the best of both the pre-modern and modern worlds, and avoids their worst features. That work can begin locally, at many different nodes, and grow upward from there.

This chapter is about building strong communities, strong in the sense that their people care about each other's well-being and about their environment. As a result of their normal day-to-day functioning, such communities would deal with major issues such as conflict, violence, poverty, hunger, and environmental pollution and depletion. Well-functioning communities would be effective not only in remedying problems when they occur, but also help prevent their ever occurring. Hunger is not likely to occur in caring communities, even if its members do not give special attention to the community's food system.

There are many levels of social organization: global, national, subnational states and provinces, cities and towns, districts, villages, neighborhoods. The focus here is on the smallest units, local communities where people live and connect with neighbors.

The emphasis here is on the process of designing caring communities, not the final result. Strong communities ought to be designed primarily by the people who will live in them. And once in them, residents should lead the redesign as their thinking and their circumstances change over time.

THE CONCEPT

Many people's lives are divided into one part in which they do things for money, and another in which they spend the money to do things they want to do. Their worlds are divided up into separate time chunks and geographical chunks. They go to some places to earn money and other places to live or play. They have distinct zones for different activities, for homes, farms, factories, and shopping malls.

The sharp separation of rural from urban through zoning laws and policies is not a good idea. It is difficult to have a real community if many residents have to be away from dawn to dusk to earn money, and spend much of their time on roadways where no one wants to be. With appropriately designed communities, more of us could do work we enjoy, near home, even if it means putting less into our financial portfolios. Many people's lives would be better if their components were more fully integrated. When people find ways to live together well, "The emphasis would shift from preoccupation with *belongings* to a more deeply satisfying focus on *belonging* "(Litfin 2010, 120).

Caring communities would be comprised of people who live close enough to one another to interact regularly. Employment opportunities, housing, and other amenities would be located in a defined contiguous space, allowing many residents to work close to where they live. These communities would have management bodies and rules determined through highly participatory processes. They would produce much of their own food, and manage energy, waste disposal and many other concerns at the community level. They would strive for sustainability, resilience, and self-reliance.

In well-designed communities, people would be more likely to care for each other and for the local environment in which they were embedded. However, the specific character of each community would depend on the character of the people who plan it and live in it. There is no single design that would suit every group of people and every configuration of the local physical environment.

Some writers believe strong communities can be created through simple measures. For example, Ela Bhatt believes that "if the six basic needs of daily life—food, clothing, housing, health, education, and banking—can largely be met with locally, within a hundred-mile radius, people will find diverse, innovative solutions to the problems of poverty, exploitation, and environmental degradation" (Bhatt 2015, 4-5). Some people think the widespread adoption of permaculture would be transformative, leading to many important benefits. Some have similar attitudes toward agroecology. The assumption here, however, is that such hoped-for outcomes are not likely unless communities are carefully designed to achieve the desired outcomes.

The 19th century economist, Henry George, was convinced that many social benefits would follow from changing tax collection procedures to "a single just charge levied upon the value of locations" (George 1879; Katzenberger no date).

Though it has had a mixed success record, the idea of intentionally designing communities has a long and honorable history. Ebenezer Howard's *Garden Cities of Tomorrow* (Howard 1902) is just one of many examples. The Vauban District developed in Freiburg, Germany offers many innovative ideas regarding not only the final design but also the planning process (Vauban 2016). The concept of Charter Cities is among the most recent community design ventures, with results yet to be assessed (Charter Cities 2012). Many ideas drawn from these efforts could be adapted and applied in other contexts. The Village Town movement offers many good suggestions (Lewenz 2007, 2011; Village Forum 2016). The communal settlements called Kibbutzim were established decades before the formal establishment of the state of Israel in 1948. Some religious communities have been sustained for centuries (Gallagher 2016). Martin Luther King's concept of the beloved community has been influential, but it has not yet been implemented in a systematic way. The Global Ecovillage Network "envisions a world of empowered citizens and communities, designing and implementing their own pathways to a sustainable future, and building bridges of hope and international solidarity" (Dregger 2016). Their annual report is full of ideas and inspiration

(GEN 2015). The Findhorn Ecovillage applies many of these ideas (Findhorn Ecovillage 2016; also see Litfin 2012).

Marinaleda is an appealing community in Spain, perhaps tarnished by some having labeled it as communist (Galtung 2016; Hancox 2013). Cohousing is emerging as an approach to balancing the need for privacy with the benefits of living together with other families. Much has been learned from community organizing in South Africa (Andersson and Richards 2016). Many resources are available through the Fellowship for Intentional Community (Intentional Communities 2016). There is a great deal to be learned from all such initiatives, the failures as well as the successes.

Historically, most community planning work has been about re-design, finding ways to improve existing communities. In some cases, the intention is to make the communities more whole in the sense described here. There are many good ways to strengthen communities, including simple actions that encourage neighborliness (Chapin 2012; Eberlein 2012a; Ipsos-Mori 2007; Itkowitz 2016). Some approaches are much more ambitious, such as Transition Towns. The Transitions movement began in Ireland, then had its ideas picked up in the United Kingdom, and the effort has stimulated comparable efforts in many other places (Transition Network 2016).

Some community design initiatives have been about creating entirely new communities in what had been sparsely occupied spaces. The exercise of planning something entirely new could be a jarring, radical experience, one that frees our thinking. It is useful to reflect on what might be possible when starting fresh, with intentionally designed physical, financial, and social arrangements. The character of the community that emerges from the planning process would depend on the views and values of the individuals who go into that planning process. If a group of like-minded people bring in all their best ideas, and have no obstacles in already-existing arrangements, they would have the potential for doing wonderful things.

In renovating existing communities, the planners could undertake some thought experiments. If they did not have differences in views with current residents or with other planning initiatives, what would they propose? If their community's physical space were to be returned back to it natural state, in their imagination, what would they propose for that space? If there were no legal constraints from outside the community, such as zoning laws, what would they do?

These exercises could help planners in an existing community get a realistic appreciation of the constraints they face. This does not mean they must submit to those constraints. The group might decide to resist or work around them in some way. They might decide not to challenge the obstacles, and instead focus on what they could do in the physical, legal, and social space they already have.

The planning dynamics would be different from what they would be for existing communities. In new communities, starting with an empty space, people get fresh choices about whether they really want to live and work together. In planning for new communities, a few leaders would call together a few like-minded people. They would be free to decide who should be invited to join them. Members who were unhappy with the way discussions were going could leave. If they found a currently unoccupied space for the new community, the planners would not have to face resistance from current residents.

Any small group could take up the challenge of creating a new caring community. One of the first tasks would be to identify a site and gain control over it. In same cases, governments might be persuaded to provide under-utilized land. Arguments could be made in terms of both social and economic benefits that would be expected. For example, it might be possible to make a convincing argument that the newly created community would help to create employment opportunities for people who otherwise depend on receiving welfare payments from governmental social support programs. Some communities might be designed specifically to relieve the burden of homelessness.

Where government-controlled land was not available, the planning group might be able to work with private landowners to learn what might be available, and on what terms. Landowners who supported the project concept might be willing to sell a site for less than full market value, especially if they wished to become part of the community or if they appreciated the social benefit that would be expected. Some nongovernmental groups such as churches might be willing to devote land to community development projects.

The planning would have to proceed on the basis of a clear understanding of the conditions that must be met to get control over the land. In some places, that control might be provided initially on a conditional basis, with the promise that if success was demonstrated after a certain time period, full control of the land would be turned over to the community.

The plan that is formulated would have to address social, physical, and financial arrangements.

SOCIAL ARRANGEMENTS

In strong caring communities, people live together well. They care about one another's well-being and treat each other well. The physical facilities could be the best imaginable, and the flow of resources might be plentiful, but if people exploit each other or ignore each other, they don't have a strong community. Every home could have its own separate photovoltaic system and its own backyard aquaponics system, but those individualized arrangements would have little to do with community-building. It is important to build the sense of community by providing many opportunities for residents to work and play together (Putnam and Feldstein 2003).

The primary function of new caring communities would be to ensure that people, individually and together, have more control over their own lives. Genuine human development is about increasing one's capacity to define, analyze, and act on one's own concerns. This means building self-reliance. In the words of Carne Ross:

If we take back agency, and bring ourselves closer to managing our affairs for ourselves, then something else may also come about: We may find a fulfillment and satisfaction, and perhaps even a meaning, which so often seems elusive in the contemporary circumstance (Ross 2012, x).

It is not only individuals but also communities that should define, analyze, and act on their own concerns (Kent 1981). Well-functioning communities are much more than random collections of people who happen to live near one another. In strong communities, there is joint effort toward a shared goal: living well together.

The community that is designed should have a good system for making its own decisions about what would serve its interests, so that it is self-governing to the extent it can be in the context in which it is embedded. Methods should be established for hearing residents' views, and for dealing with conflicts within the community. There could be town hall style meetings in which all interested parties could participate in discussing issues of concern to the community. Details regarding decision-making procedures could be spelled out in the community's charter and bylaws.

The emphasis should be on community self-reliance, which emphasizes local control, rather than community self-sufficiency, which refers to local production of goods to meet local needs. Self-reliance allows for trade and other kinds of interactions with others according to the community's best judgment about what would benefit its members and their environment (Kent 2011, 124-132).

The community-centered way of living would support not only self-reliance but also the principle of subsidiarity, the idea that "each social and political group should help smaller or more local ones accomplish their respective ends without, however, arrogating those tasks to itself" (Carozza 2003, 38, note 1; also see Bosnich 1996; Minus 2004). To the extent feasible, how you live should be decided in your family and your community, not in higher levels of government.

Local communities should be involved in disaster planning, and not assume that distant government agencies should do all the planning that is needed. No one is more interested in serving the community's interests than the community itself. Outsiders always have other priorities.

In new communities, it would be possible to set up clear procedures for becoming a resident of the community, and also for leaving it. Entry could be based on a combination of commitments to pay to enter (as one might pay to enter a retirement community or a cruise ship), commitments to provide services to the community, and commitments to respect certain values. The contractual agreements could have some fixed elements, common to all, but could also have some elements that are individually negotiated. The community's managers should be open to creative proposals for entry.

Communities could be set up as rights-based social systems, governed by their charters and bylaws, under rules shaped and endlessly refined by their participants. Unlike the global human rights system, in which rights appear to be formulated and interpreted by distant others, caring communities would involve rights, obligations, and systems of accountability that are formulated locally, with full participation of the residents.

Social innovations could be introduced not only in the community that is created but also in the planning process leading up to its creation. To illustrate, a group in New Zealand, grounded in the indigenous Maori culture, developed Tipu Ake, "a leadership model that can help us see organizations, teams and individuals as living organisms growing as part of a complex ecosystem" (Goldsbury 2010, 3). Frances Moore Lappé calls for a comparable ecology-oriented way of thinking (Lappé 2011a), as do Elinor Ostrom (Ostrom 1990) and Elisabet Sahtouris (Lifeweb 2016). Such innovative approaches to planning could be adapted to the governance of the community itself.

Each new caring community that is designed should be viewed as just one in a variety of such communities. They would have some features in common, but each of them would also have its own distinctive character, depending on who designs it and who lives in it. At one site the community might be bound together by the participants' interest in cultural preservation, in another by their interest in agroecology, and in another it might be their shared religious faith. The various sites would be diverse, and accommodate many different people through that diversity.

PHYSICAL ARRANGEMENTS

The designers of a new caring community would have to determine the characteristics of the site in terms of the soil, terrain, water availability, sewage services, electricity supply, wind patterns, etc. The zoning constraints would have to be clarified, to determine what activities and buildings are permissible on the site. Judgments would have to be made about whether some zoning variances might be feasible.

Designs for housing could adopt the best available environmental technologies, and the buildings could be laid out in ways that would facilitate social interaction. Options could include small clusters of homes within the larger community, in "pocket neighborhoods" (Abrahams 2012) while others might prefer medium- or high-rise apartments with various arrangements of amenities in and around the buildings.

A multi-purpose community center would be essential. There should be good playgrounds for children, and park and picnic spaces. Walking and bicycling trails should be provided.

Facilities would be provided not only for housing but also for employment opportunities. There could be a community farm, managed as a cooperative by those residents who wished to participate in it. The housing units, many with their own gardens, could be arrayed around the farm. There could be retail shops and restaurants. There could be a small workshop fitted out with basic tools. There could be a commercial kitchen available for small-scale food processing, teaching, and preparation of festival meals.

Provisions would have to be made for all the services that are required, such as schools, health services, and places of worship. Full consideration would have to be given to their physical, social, and, and financial aspects. While many services could be provided from within the community, some, such as airports, higher education, and high level medical care, could be accessed from outside.

Mobility within the community could be based on walking, bicycles, and small electric vehicles. For trips outside the community, bus services might be available. Cars could be privately owned or available for rent on an hourly or daily basis.

Energy production and consumption could be managed at the community level, with the housing units, the farm, and the various businesses drawing much of their electricity supply from shared solar and wind systems. Buildings and other facilities would be designed to be energy efficient. Community-wide waste management could be designed to minimize the need for shipping anything to land-fills (Eberlein 2012b).

Many people have had experience in sharing ownership and sharing benefits in commonplace arrangements such as community gardens, libraries, and parks. The ideas can be adapted to other sorts of amenities. For example, modern telecommunications and solar gardens could be organized in ways that help strengthen the local sense of community (Bezdomny 2015; Solar Gardens 2016).

Physical arrangements should be economically and environmentally sensible, and where feasible, they also should facilitate social interaction and thus help to build the sense of community, the caring.

FINANCIAL ARRANGEMENTS

The finances would have to be worked out, either for starting up a new community and operating it over time, or to cover the incremental changes proposed for an existing community. Just as the physical arrangements should support the social objectives, the financial arrangements also should support the social objectives.

Whether by choice or necessity, many people now spend a great deal of money on their homes. Depending on the means available to them, people shop not just for shelter, but also for a variety of amenities. The designers of a new caring community would anticipate what people would value in terms of lifestyle, employment, housing, social networks, and other aspects of their lives.

Some employment opportunities could be in pre-determined salaried positions. Some people could come in with their own proposals, saying they would like to work as artisans or trades people of various kinds: bakers, plumbers, carpenters, artists, and musicians. The contracts could take various forms, just as the participants on a large cruise ship all have agreements of different kinds about the terms on which they participate.

The cruise ship metaphor has an important quality: everyone on board makes a deliberate choice to be there. No one is there as a result of a historical accident or inertia. A community that is newly designed would have that quality, with a few exceptions for people who had already lived on the site and were accommodated in the plan.

Of course the social life on cruise ships could be designed in many different ways, ranging from highly egalitarian to the worst sort of caste system (Schwartz 2016). One key to limiting abuse is to make sure that getting on is optional and the option of getting off is always available. Ensuring easy exit would place a limit on dissatisfaction. Unhappy people would leave.

Financing to create a community designed primarily for high-income residents would be relatively easy to obtain. Financing a community for people with low incomes would be more challenging, but not impossible. People can be hugely resourceful, and accomplish a great deal even on what others might regard as useless land. Many communities produce much of their food even on rocky, mountainous, or arid lands. The potentials for producing food under difficult conditions are demonstrated by the ways people have established productive gardens even in wartime (Helphand 2006).

There is a risk that communities designed by and for low-income people could turn into slums. In some contexts, a better solution would be to mix residents of different income levels. The design could accommodate entry into the community under various financial arrangements, possibly involving some cross-subsidy to ensure that the community works well for all who participate. In some contexts, communities for poor people could be successful if close attention was given to their social organization (Green 2016).

The new community would be a good place for trying out innovative business models such as those advocated by BALLE, the Business Alliance for Local Living Economies BALLE 2016). It shows how locally owned and operated businesses can serve the communities in which they operate

The community could create its own financial institutions. It could have its own credit union. It could include a micro-loan facility that would help arrange group support for borrowers, as envisioned in the original model of the Grameen bank in Bangladesh:

> *The Grameen Bank is based on the voluntary formation of small groups of five people to provide mutual, morally binding group guarantees in lieu of the collateral required by conventional banks. At first only two members of a group are allowed to apply for a loan. Depending on their performance in repayment the next two borrowers can then apply and, subsequently, the fifth member as well (Grameen Bank 2010).*

The community could also establish various forms of safety nets for participants who get into financial or other difficulties, to help get people back on their feet.

The community as a whole could be organized as a corporation, a condominium, a limited partnership, a cooperative (Baarda 2006; Duda 2016; International Year of Cooperatives 2016), or some other type of enterprise. Ideas for worker-owned corporations (Kelly 2012) could be adapted to form resident-owned corporations.

Another approach would be to create a community land trust (CLT) (Axel-Lute 2012; Diacon 2005; Lopez Community Land Trust 2016; The Biodynamic Land Trust 2016). The town of Letchworth in the United Kingdom has operated as a CLT for more than a hundred years, drawing on many of Ebenezer Howard's ideas on urban agriculture. All the residents are lessees from the trust (Letchworth Garden City 2016). Burlington, Vermont has a highly developed CLT, described by one observer as follows:

> *Usually CLTs manage land for Housing and Economic activities. However, in Burlington, the model has expanded toward the man-*

agement and the provision of urban land to local farmers that are feeding the city and beyond. The Intervale Centre which is run as a trust, leases or buys land from the government to allow small farmers to start their business at individual or collective level (Intervale Center 2016). Once the farmers start making profits (usually 3 years or even more) they will start paying back for the use of the land. The central consideration from a legal point of view is the separation between the freehold (property right) on the land, that remains in the hands of the trust, and the leasehold that can be either individual, or collective, for instance in the hands of cooperatives (Cabannes 2012, 29-30).*

Public-Private Partnerships are another possible model:

Public Private Partnership (PPP) is a cooperative approach between the public and private sectors. Such partnership mechanisms have been widely used around the world to promote development in areas of infrastructure, public utilities and services, etc. Because of the lack of access to capital and/or expertise, the PPP may provide not only the most effective but perhaps the only way to realize a socially necessary project.

However, though attractive in theory, PPPs require a very well thought out approach on the part of all the partners, careful drafting of contracts and assurance that not only the economics are considered but the reality, needs and aspirations of all the stakeholders. In some cases such as in "Build-Operate-Transfer (BOT)" projects, a Public Private Partnership can be seen as a socially responsible investment where enterprises have the opportunity to obtain a return on investment and contribute to society. (Santa Barbara 2009, 141)

China has used this approach in its Guangcai Model, "a multi-sector partnership between entrepreneurs, governments, nongovernmental organizations and farmers to reduce poverty in rural areas" (Santa Barbara 2009, 141).

Whether it is government or private investors that provide the startup capital for establishing a new community, the monetary return on investment is not likely to be as high as might be obtained in other kinds of ventures. However, there would be an expectation of a significant non-monetary return, whether in terms of alleviation of poverty or other social objectives.

Design and construction of some of the community's facilities could be left to the community itself, thus reducing startup costs. The community could even fabricate some of its own construction machines (Open Source Ecology 2016).

Understanding that things could go bad, there should be orderly procedures ready for dismantling the community if that should become necessary. If it must die, it should die with dignity.

In most cases the design should require that once the community enters its operational phase it should be economically self-sustaining. However, individual residents would be welcome to bring in income, loans, grants or subsidies for themselves, and they would be free to spend their money inside or outside the community.

Any group that has a shared vision regarding a community of the sort described here could pursue it together. That vision might be blurry at the outset, but the group members could move toward a sharper, shared focus by exchanging ideas and sketching out possibilities in a series of meetings. When they felt ready, they could continue on by jointly preparing a proposal for professional planning support and start-up of the community.

There are many potential funding sources, locally, nationally, and globally. Some sources might be private, such as foundations, and some might be governmental.

There are funding sources other than grants. For example, the Social Finance organization "has been expanding the definition of social finance by providing loans, gifts, and investments that foster social and spiritual renewal (RSF Social Finance 2012)." Kiva facilitates lending and borrowing for small projects to alleviate poverty (Kiva 2016). Ioby "brings environmental projects to life" (Ioby 2016). In many places there are similar local funding sources. There are many sources of "crowdfunding" for small projects (Steinberg 2012). The Grassroots Institute for Fundraising Training offers training and tips (Grassroots 2016). New approaches to obtaining and to thinking about money are offered by the Money Alliance (Slow Money 2016).

ECONOMICS

The preceding section, on finances, is about money. Here we take a broader, more philosophical view, discussing what should be valued and how relationships should be organized. Conventional economic thinking is based on the assumption that people are mainly concerned about their own well-being and give little attention to the well-being of other people or the well-being of the environment in which they are embedded. It is based on relationships of indifference and exploitation. In this book, however, we explore the idea of economics based on caring. This approach is well developed in Riane Eisler's book, *The Real Wealth of Nations: Creating a Caring Economics*. Caring economics are grounded in *partnership* or mutual respect systems, rather than *domination* or top-down control systems. The basic premise is that "economic systems should promote human welfare and human happiness" (Eisler 2007, 3, 4; 2016).

The discussion here draws on Eisler's thinking, but acknowledges that it may be too ambitious to hope for large-scale conversion from the dominant old kind of economic thinking to this new kind. The global economy and most national economies are too big and too set in their ways. There are alternatives (Cavanagh and Mander, eds. 2004; Lindner 2012).

We cannot realistically expect the deep caring that is needed will span great distances. People are much more likely to care about their neighbors than about distant strangers. Thus, where Eisler contemplated the real wealth of nations, the premise here is that it would be more feasible to focus, at least initially, on the real wealth of local communities.

Eisler calls for a radical reformulation of economics, and for more explicit valuing of caring activities such as those undertaken by parents in looking after their children. The view here, however, is that deep caring will not follow from the reformulation of economic concepts and accounting methods. Rather, the reformulation of economics will follow from deep caring. That deep caring can be found and cultivated in strong communities.

THE DESIGN PROCESS

While it is useful to distinguish among social, physical, and financial/economic aspects of the community, they really cannot be neatly separated into different compartments. They must be designed together, simultaneously. This weaving together of diverse ideas is well illustrated by the concept of the Food Commons (Food Commons 2016), which is really about many issues, not just food issues. Designing a caring community requires holistic thinking. There are tools available to support that work (Binswanger-Mkhize, De Regt, and Spector. 2010; Building Living Neighborhoods 2016; Lewenz 2007; Lewenz 2011; Pattern Language 2016; REconomy Project 2016).

Many elements of the design could be encapsulated in contracts between the community-as-a-whole and individuals or families that wish to join it. It could spell out details of the social, physical, and financial/economic arrangements as they relate to these particular individuals. The contract could be comprised of a basic template, with some elements left open for negotiation. The template itself would be revised from time to time, depending on what is learned.

Any number of different groups could plan communities of the sort described here. Several design processes could be carried out in parallel, for different sites, with exchanges of insights among them. They would have in common their recognition that, while there are many wonderful ideas about how to do the social, physical, and financial things in better ways, so far the applications of those ideas have been very scattered. Those ideas could be pulled together in diverse new and renovated communities.

Chapter Eight
Caring About Local Hunger

CHANGE FROM THE TOP?

Many observers concerned about hunger criticize the global food system and call for it to be replaced (Field and Bell 2013; Hines 2004; Other Worlds 2013; RTFN-Watch 2012). The United Nations Conference on Trade and Development calls on global agriculture to "wake up before it is too late" (UNCTAD 2013). Many people hope that national and global agencies will address humanity's great challenges such as conflict, poverty, hunger, pollution, and resource depletion. They have not been very successful. It is time to stop looking upward for solutions.

Nobel laureate Elinor Ostrom argued that a single international agreement at the Rio+20 environmental summit in June 2012 would have been a grave mistake:

> *We cannot rely on singular global policies to solve the problem of managing our common resources: the oceans, atmosphere, forests, waterways, and rich diversity of life that combine to create the right conditions for life, including seven billion humans, to thrive. . . . This grassroots diversity in "green policymaking" makes economic sense. "Sustainable cities" attract the creative, educated people who want to live in a pollution-free, modern urban environment that suits their lifestyles. This is where future growth lies. Like upgrading a mobile phone, when people see the benefits, they will discard old models in a flash. . . . (Ostrom 2012)*

There is little hope for solving food and nutrition problems globally if we don't know how to address them locally. Radical transformation from the top is not likely. Global caring is shallow, spread too thinly, and it is usually seen as unilateral, top-down. It is about the haves doing things for the have-nots. Deep mutual caring is more likely in local communities, neighbor to neighbor. In working to end hunger and many other miseries, it makes sense to view communities as the fundamental unit of analysis and locus of action.

COMMUNITY-BASED APPROACH

Statistical descriptions of hunger don't get at the roots of the problem. Studies of factors causing hunger (e.g., FAO 2013b; IFPRI 2013, 102-120; Roser 2016a) rarely cover social relationships and how people actually live and relate to one another. The World Food Programme says the most important causes of hunger are the poverty trap, lack of investment in agriculture, climate and weather, war and displacement, and unstable markets (World Food Programme 2016). Indifference and exploitation are not mentioned. The WFP is regularly disappointed by the weak responses to its appeals for funding to deal with the world's most extreme hunger situations, but this weak response cannot be explained by the six factors in its list.

Analysts may speak about deficits in land availability, water, seeds, knowledge, and trade opportunities, but fail to see that the major problem might be a deficit in caring. They do not appreciate that at its root, hunger is a social problem, heavily influenced by human relationships of caring, indifference, and exploitation.

Hunger arises when people don't have adequate control over their own life circumstances. Where people go hungry, we can be sure that others are controlling the resources around them and shaping the terms on which they live. The others are serving their own interests, not those of the hungry. People need power, individually and in community with others, to shape their own lives and live in dignity.

In strong communities, where people care about one another's well-being and about the environment in which they are embedded, no one goes hungry. This is true even in poor and in so-called primitive societies. Karl Polanyi recognized this:

It is the absence of the threat of individual starvation which makes primitive society, in a sense, more humane than market economy, and at the same time less economic. . . . . [A]s a rule, the individual in primitive society is not threatened by starvation unless the community as a whole is in a like predicament. . . . destitution is impossible: whosoever needs assistance receives it unquestioningly. . . . There is no starvation in societies living on the subsistence margin. (Polanyi 1944, 171-172)

George Kanahele said the same thing about pre-contact Hawai'i:

The starkest forms of famine occur in much more harsh natural environments than Hawai'i's and, ironically, in part as a result of the industrialism which makes marginal economies dependent upon international political and economic events over which people in such economies have no control. We cannot honestly imagine absolute hunger occurring among the families dwelling in a self-sufficient 'iliahupua'a in the days of old. (Kanahele 1986, 324)

Others put it this way:

When a community functions well, it is because of the active solidarity among its members. People look out for each other, help each other . . . When individuals slip into poverty it is not simply because they have run out of money—it is also because their community has failed. (Dessewfy and Hammer 1995)

Strong communities are the ultimate safety net. This is well illustrated by the state of Kerala in India. It has more equitable distribution of food between income groups and within families and better access to and utilization of health care despite its being poorer than most other states in India (Banik 2007, 4). Kerala's people live together well on every dimension, not just in relation to food and health issues (Nair 2016).

Kerala's state government formulates and implements its social policy in close consultation with the poor (Banik 2011, 96), clearly indicating the government's care for the poor. This contrasts with the states in India that show "disdain for the poor" or "seldom attempt to monitor the nutritional status of vulnerable groups" or where "the entire administrative response is geared toward denying allegations of starvations deaths . . ." (Banik 2011, 99). There is no easy way to measure caring, but sometimes its presence or absence is obvious.

There can be serious food supply issues when geophysical hazards such as earthquakes and floods occur, or when armed attacks suddenly disrupt local food systems and entire communities. However, in stable communities, hunger usually results from exploitation, where some people profit excessively from the fruits of other people's labor. When people have decent opportunities and can enjoy the full benefits of their own labor, they live adequately. They do that even in harsh physical environments. Where physical and social environments are too harsh to sustain life, people try to move elsewhere. In exceptional circumstances, hunger can occur anywhere. But even a brief survey of stable intentional communities throughout the world (Dregger 2016) would lead to the confident belief that in those places no one goes hungry.

In many high-income countries, there are low-income groups that go hungry. Their problems may be due as much to the absence of caring communities as to the lack of money. In Japan, for example, where increasing numbers of senior citizens are arrested for shoplifting . . .

> *"Senior citizens shoplift lunch boxes and bread out of poverty, and they also steal because they are lonely and isolated"* . . . . *Some steal even when they aren't really hungry because the traditional support system is breaking down and they have become isolated from society . . ." (Nohara and Sharp 2013)*

Society's indifference takes a heavy toll on the isolated elderly.

When economic gaps widen sharply, as described in Chapter Four, governments could arrange programs to compensate for the gap-widening effects of ordinary economic behavior. Failing to do that shows there is not enough caring about those who are at the wrong end of the squeeze of the middle class.

Commercial food production tends to serve middle- and high-income people because it is designed primarily to produce good incomes for the owners. Government agencies tend to favor the same middle- and high-income people. People with low incomes and little political power often get their food outside the dominant commercial system, by producing food themselves in subsistence farms or backyard gardens, by purchasing from small-scale farmers who have little access to major markets, and by cooperative efforts such as community gardens.

FOOD PROJECTS

As argued in the preceding chapter, caring can be strengthened by encouraging community members to spend more time working and playing together. That would reduce the likelihood that anyone in the community goes hungry. The likelihood would be even smaller when the joint activity is about food. The sense of community can lead to many different food projects, and those projects in turn can help to build the sense of community (Brown 2013). They would show the presence of caring and also help to strengthen the caring.

There are many options. For example, farms could be organized as collective community-based enterprises. People could garden together, cook together, and eat together in many different settings. Food-related skills could be strengthened through the sharing of knowledge and hands-on experience. People who are facing difficulties could be offered food packages or meals, and could also be given support in learning how to grow food, shop better, and cook for themselves (Pascual and Powers 2012). Community farms could raise food products mainly to meet their own needs, but at the same time produce a few specialty products for sale outside the community (Sanz-Cañada 2016).

Communities could establish local food policy councils to be permanently attentive to local food and nutrition issues (Burgan and

Winne 2012; FAO 2011; Food Policy Networks 2016; Kent 2011, 142-153). One good way to stimulate serious reflection by local people on the local food situation is to facilitate studies of it by them (CUNY School of Public Health 2016). If people measure and chart their own children's height and weight, they would be likely to gain a good understanding of their children's stunting and wasting, and likely to act on the problems.

Ideas for creating a "food commons" might be applied within the community, through cooperative arrangements among farmers, processors, wholesalers, retailers, and consumers (Food Commons 2012). Food procurement policies of schools, hospitals and other institutions could be designed to favor food producers within the community.

There are many opportunities for creativity. In Chicago, local groups are addressing the problem of food desserts by selling fruits and vegetables from a bus that makes regular visits to the neighborhoods (Jennings 2013). Communities can sell their produce to outsiders through a scheduled farmers' market or through trucks that sell the produce outside the community (Veggie Mobile 2016).

Most food is distributed through conventional marketing methods, but people in caring communities are also likely to share their gardens' produce with their neighbors, or they might share jams, breads, and cakes. Sharing of this sort can be carried to surprising extremes. In the British town of Todmorden, for example, people raise fruits and vegetables and invite others to harvest them even without asking (Graff 2011; Incredible Edible 2013; Warhurst 2012). Some small farms inside the city limits of Detroit supply vegetables for anyone who wants them, there for the taking (Urban Roots 2012). In South Central Los Angeles, vegetable gardens are being placed in abandoned lots and traffic medians (Finley 2013). Small local seed libraries can be set up to help people start their own home gardens (Steiberger and Peterson 2016). If it is well managed, the sharing of breastmilk through direct contacts or through human milk banks can be of great benefit not only for health but also for building the sense of community (Davies 2016).

Food sharing is commonplace, especially in low-income communities (Morton et al. 2008). It can be enhanced in many ways, through community festivals and pot-luck meals, perhaps on the basis of a regular schedule. Soup kitchens of various forms could be established (Bayne 2013).

In addition to sharing food, there are always opportunities for sharing information, advice, and ideas. Gardeners frequently engage in sharing of this sort. Dietary advice can be shared. The dietary guidelines issued by the Brazilian government in 2014 (Ministry of Health of Brazil 2014; Monteiro et al. 2015) might be of particular interest. They could be adapted to fit other contexts.

The Shareable website offers ideas for creative sharing, including many centered on food (Shareable 2016). There are creative ways to facilitate meal sharing (Johnson 2013). The nongovernmental organization Heifer International promotes sharing systematically through Passing on the Gift, a program in which low-income people who receive donated animals "share the offspring of their animals – along with their knowledge, resources, and skills – an expanding network of hope, dignity and self-reliance" (Heifer International 2013). The sharing of mothers' milk is now supported in systematic ways (PATH 2013; HMBNA 2016).

Sharing can be facilitated by having people set up tables at farmers' markets to accept people's excess fruits and vegetables and give them to other people who need them (Cave 2013). Such transactions have been studied as the *gift economy*, in contrast with the conventional *exchange economy*. Many pre-modern food systems use non-market modes of exchange that are not very visible. They are beyond the comprehension of modern neo-classical economics, but they can work very well (Sahlins 1972).

There are many organizations that encourage gardening, including some, such as Gardens for Health International, that promote gardening specifically to prevent malnutrition (Gardens for Health International 2016). While the explicit focus for gardening projects may be on the food that is produced, gardening can also help to build stronger communities (Schukoske 1999). The American Community Gardening Association is devoted to the idea of building community through gardening (American Community Gardening Association 2016).

Farms, markets, and restaurants could be set up as cooperatives of various forms. There are organizations that can advise on how to set up cooperatives in harmony with local cultural practices and traditions (Kohala Center 2011). Northern Italy demonstrates the benefits of having entire regions organize their businesses as cooperatives (Luna 2013).

Community members can be encouraged to take their meals together in various settings, to enjoy each other's companionship. They could also join together in the food preparation work and the cleanup following the meal.

In some places groups of friends take turns hosting meals at their homes in a more or less regular cycle. Some cohousing communities organize frequent common meals (Blank 2014; Villines 2014). Block parties and fiestas also help to build the sense of community.

One observer speaks of "the central role of conviviality: the pleasure of sharing food with others, of celebrating communal culinary traditions and life at large":

> *The word 'convivial' derives from Latin, where it means quite simply 'the act of living together.' We are drawn to conviviality by our need for safety, companionship and comfort. But in today's hyper-efficient, fast-paced world, we often sacrifice that which made us human—our fundamental need for food—and the communality that was born of this need. Instead, we rush from one task to the next and eating becomes just another chore to be slotted into our busy schedules. (Middelmann-Whitney 2010; also see Gopnick 2012)*

Convivial meals contribute to the nourishment of bodies and also to the nourishment of communities. An observer with Slow Food USA articulates the main point:

Imagine a world in which we can no longer point to the paradox of skyrocketing obesity and skyrocketing hunger as a symptom of injustice, and can instead point to the disappearance of both as a symptom of justice flourishing. I believe the best way to begin is through building meaningful human relationships, through linking people and communities together around a sense of common purpose. Groups of people become communities by sharing work, sharing struggles, and sharing food. This leads to real, personal relationships; a sense of co-dependence and co-commitment. Once you've shared a meal with someone, or worked on a project together, you view each other differently. You're more likely to take care of each other, and, I believe, you're more likely to stand together and work for change together. (Viertel 2012)

Brazil's dietary guidelines now recommend, "Eat in company whenever possible" (Tsai 2016).

TECHNOLOGY

Communities should be innovative in the organization of their economies, and they should judiciously take advantage of both premodern and modern technologies. For food production, they should consider options such as permaculture and aquaponics. People who have no access to land can grow things on roofs and walls and in bags.

There are innovative methods for harvesting water and energy, and for growing things with little water and little energy. Methods have been developed for dry farming and for turning dry areas into fertile farmland (Sutton-Redner 2014). There are ways to graze livestock on land that others view as useless, and through that means, "move massive amounts of carbon and water from the atmosphere back to the soil and begin reversing thousands of years of human-caused desertification" (Savory Institute 2016). Such approaches are important when trying to create new communities for people with low-incomes since they are likely to have access only to what others regard as low-value land.

While such technologies can be helpful, they can also be disappointing. There was an excess of optimism associated with the appropriate technology movement that followed publication of E. F Schumacher's *Small is Beautiful* (Schumacher 1973). Some failures result from trying new hardware in old social arrangements.

Innovations such as permaculture or aquaponics are good in environmental and economic terms for the families that adopt those techniques, but they may do little to build the sense of community. When operated by groups, however, they can help to build that sense.

The Internet and its associated tools now can help food producers in technical ways, and they can also facilitate social relationships, including those relating to food (Shareable 2016). A good video on, say, how to grow and prepare orange-flesh sweet potatoes, accompanied by Internet-based discussions could help to improve people's nutrition in many different places at a negligible cost. With the help of the Internet, willing advisers can offer support from anywhere in the world.

COMMUNITY SELF-RELIANCE

Many people call for increased local self-sufficiency based on producing rather than importing food. They feel it is important to replace imports with locally produced food, but the reasons are not always clear. Pushed too far, the import-substitution strategy could lead to higher food costs, and it could reduce the diversity in available foods. Local is often better but it is not always better.

It is important to distinguish between *self-sufficiency*, which refers to local production to meet local needs, and *self-reliance*, which emphasizes local control, but allows for exchange with outsiders. Increasing self-sufficiency to some extent may be wise in some circumstances, but the more fundamental need is to increase self-reliance. Communities should be able to make their own decisions about how they provide for themselves.

Self-reliance means giving mindful attention to relationships with others. It requires an open and democratic decision-making body, perhaps in the form of a well-organized food policy council. Importing and exporting food to and from a locality is fine so long as local people have made a fair and informed judgment about what serves their interests.

In any community, genuine development means the increasing capacity to define, analyze and act on one's own problems. Development, properly understood, is about building self-reliance, not about accumulating money. Self-reliance is about responsible autonomy.

The decisions that are made should consider not only the well-being of people inside the community, but also the impacts on people outside. Importing foods from low-income areas is good if it helps to improve the well-being of people in distant places. It is not good if it promotes their exploitation.

Consideration for the well-being of outsiders means that advocates of urban agriculture should consider its impacts on the nearby rural people who have traditionally supplied the cities. If cities provide for themselves, what are farmers in rural areas to do for a living?

In many places, higher levels of governance impose their approaches to food production on local communities. The alternative is for communities to take the lead, exercising at least a degree of local food sovereignty (Anderson, Pimbert, and Kiss 2015; Pimbert 2009). Community leaders need to figure out what makes sense at the local level, and then call on the higher levels of governance to support that model. Under the principle of subsidiarity, the primary task for agencies at every level in the rings of responsibility should be to do what is best at the local level, for and with the people on the ground.

Designers of food systems for local communities should study the Milan Urban Food Policy Pact (Milan Urban Food Policy Pact 2016). Cities can be understood as collections of small diverse communities in their various neighborhoods. Each of them has to relate to the food policy established by the city and higher levels of governance. Each must adapt to their contexts and participate actively in the shaping of food policy.

WEALTH OF STRONG COMMUNITIES

Community-based food production is based on the recognition that while participants might have little cash income, they have other kinds of wealth such as their labor power, their motivation, and their knowledge of the local culture and the local environment. There is also natural wealth in the local land, water, and sunshine that can be used in sustainable ways. But perhaps most important, in strong communities, members care about one another, an important asset (McKibben 2008; Pilisuk and Parks 1986).

The inputs to community-based food operations are different from those used by commercial ones, and their managers are likely to have different priorities regarding what are the important outputs. With their unconventional economics, community-based food operations might be feasible even where commercial operations are not.

People with little money can live together with no one going hungry, as demonstrated in countless places over thousands of years. Instead of focusing on ways to remedy hunger when it occurs, we can devise ways of living in which the hunger issue does not come up (Dregger 2016).

Hunger in the world is not due to a lack of knowledge about nutrition or a global shortage of material resources. There are local shortages of various kinds, but not global shortages. The local shortages are rooted in conflicts of interest about how the earth's abundant resources should be used. Hunger persists because many people don't care much about others and many people exploit others.

Food systems are social as well as technological, establishing specific relationships among people in the community. Well-designed food systems reflect and strengthen positive, caring relationships among the people, and through that means help to ensure the food security and general well-being of the entire community. The caring that is built up through collaborative food projects is likely to yield many benefits beyond the food sector itself.

If the linkage between nutrition and agriculture is to be restored, it will have to be done at ground level, in the communities. High-level agencies could offer important support services, but the main action would have to be local, in the communities. The restoration of that linkage would come not from market forces but from the fact that people care about each other's well being. If all communities' food systems were designed to ensure that their people were well nourished, we would have a world without hunger.

The argument here can be summarized in three propositions:

- Hunger is less likely to occur where people care about one another's well being.

- Caring behavior is strengthened when people work and play together.

- Hunger in any community is likely to be reduced by encouraging its people to work and play together, especially in food-related activities.

Many agencies, both governmental and nongovernmental, do things to improve the nutrition situation. The World Health Organization has developed excellent guidelines for dealing with specific nutrition problems (World Health Organization 2016). These are important building blocks, but we do not yet have a clear vision of the building. It is difficult to find any agencies that have imagined a world in which there is no hunger, perhaps comparable to a world in which there is no polio. If you cannot envision something, you cannot build it. Even a chair cannot be built without first imagining it.

In *The Conquest of Bread*, Peter Kropotkin argued, "Well-Being for all is not a dream. It is possible, realizable, owing to all that our ancestors have done to increase our powers of production (Kropotkin 1906)". There are no technological obstacles. Whether we actually seek well-being for all is another question. Collectively, we have the capacity to end hunger and other miseries in the world. Whether we have the motivation—the caring—to actually do that is a different issue.

All hunger is local and must be addressed locally. This can be done through strong, caring communities. The proper functions of agencies at the higher levels of governance is to support individuals and agencies at the local level, in the communities, to ensure that the families and neighbors they care about have their needs fulfilled. Everyone should have the opportunity to live with others in a strong caring community.

Creating caring communities is a plausible pathway to addressing many different kinds of human misery. Strong communities serve as good protection not only from hunger but also from poverty, crime, violence, pollution, and many other negative aspects of modern life. Strong communities can provide remedies when such problems occur, but more importantly, they can prevent the problems from ever occurring.

Hunger will not end as a result of missions sent to feed the hungry, but by finding ways of living together that prevent hunger and other miseries from ever occurring. Many individuals and agencies have done hugely important work in rescuing people from hunger, but the *prevention* of hunger will come from the regular day to day concern of people for their neighbors.

In her visionary study, Riane Eisler described caring economics as the basis for the real wealth of nations (Eisler 2007). The concept did not take hold, perhaps because nations are too large and internally divided for strong caring. Caring economics makes more sense in small caring communities.

In strong communities, where people care about each other's well-being, people don't go hungry. That would be expected even if the community did not pay special attention to its food system. But if communities did build up their control over their own food supplies, the impact would be even greater. This is clear to Professor Olivier de Schutter, the distinguished former United Nations Special Rapporteur on the Right to Food:

The more I have worked with governments operating from the top down, the more I have come to believe in the strength of social movements to make change happen from the bottom up. Solutions that can be designed using local resources (in addition to, not instead of, external resources that may provide backup) are less vulnerable to outside market or energy shocks. The more diverse these solutions, the better local systems will be equipped to deal with contingencies.

Is this revolutionary? Perhaps not if we think of a revolution as an event in history when a group overthrows a regime and takes power. That view of revolution however, as German political philosopher Hannah Arendt once remarked, sounds more like a coup d'état. Changing society without seizing power is what food-sovereignty movements are about. The revolution they propose is a silent one. It is gradual. But it is already happening all around us, proposing an alternative to low-cost, big-food systems with which we've been saddled for far too long. (De Schutter 2015)

Remarkably, discussions about how to deal with the hunger issue have focused on food production and have given little attention to the role of human relations. To end hunger, we will first have to show how to live together well locally. We need to get beyond talking about how we ought to live and actually demonstrate it. Others will emulate the successes. When we find ways to live together so well that no one goes hungry, we will discover that living in a caring community is itself nourishing and a form of wealth.

References

3 News. 2012. "Fonterra Hopes Milk Campaign Will Drive Demand." *3 News.* December 14 http://www.3news.co.nz/Fonterra-hopes-milk-campaign-will-drive-demand/tabid/421/articleID/280345/Default.aspx#ixzz2EyezfXwk

Abrahams, Sally. 2012. "Share Common Ground." *AARP Bulletin.* May 7. http://www.aarp.org/home-family/livable-communities/info-05-2012/pocket-neighborhoods-common-ground.html?cmp=NLC-WBLTR-PMCTRL-052512-F3-13&USEG_ID=

ABC TV 2013. "Lions, Tiger and Bears living Together" *ABC News.* http://abcnews.go.com/WNT/video/lions-tigers-bears-living-18928105

ACDI-VOCA. 2016. *Nutrition-Sensitive Agriculture Increases Food Consumption and Dietary Diversity.* ACDA-VOCA. April 14. http://acdivoca.org/resources/newsroom/news/nutrition-sensitive-agriculture-increases-food-consumption-and-dietary

Alston, Philip, and Katarina Tomaševski, eds. 1984. *The Right to Food.* Dordrecht, The Netherlands: Martinus Nijhoff. http://books.google.com/books?id=Z51Of2mF_f4C&pg=PR1&lpg=PR1&dq=right+to+food&ots=EvzYmmC-tU&sig=Jkqz9J1URn10Ca2TYevKHsMiPGg&hl=en#v=onepage&q=&f=false

American Community Gardening Association. 2016. Website. http://www.communitygarden.org/about-acga/

Anderson, Colin, Michel Pimbert, and Csilla Kiss. 2015. *Building, Defending and Strengthening Agroecology: A Global Struggle for Food Sovereignty.* United Kingdom: Coventry University Centre for Agroecology, Water and Resilience. http://www.agriculturesnetwork.org/news/bulding-defending-strengthening-agroecology-25092015

Andersson, Gavin, and Howard Richards. 2016. *Unbounded Organizing in Community*. Dignity Press.

Axel-Lute, Miriam, John Emmeus Davis, and Harold Simon. 2012. "Cheaper Together: How Neighbors Invest in Community." *Yes! Magazine* (May 1). http://www.yesmagazine.org/issues/making-it-home/cheaper-together-how-neighbors-invest-in-community

Axelrod, Robert M. 2006. *The Evolution of Cooperation*. Revised Edition. New York: Basic Books.

Azizi, Arash. 2009. "Supposed Hafiz Poem Recited by McGuinty Turns Out to be Fake." *Payvand Iran News* (April 23). http://www.payvand.com/news/09/apr/1266.html

Baarda, James R. 2006. *Current Issues in Cooperative Finance and Governance*. Washington, D.C.: United States Department of Agriculture. http://uwcc.wisc.edu/info/governance/baard.pdf

Baby Milk Action. 2010. *The Resource Centre*. http://www.babymilkaction.org/resources/yqsanswered/yqanestle09.html

Baby Milk Action. 2016. *Trade vs Health—WHO Opens the Door to Big Business While Trying to Protect Babies*. United Kingdom: Baby Milk Action. http://www.babymilkaction.org/archives/9786

Baker, Phillip. 2016. "Is an Infant Feeding Transition Underway?" *PLOS Blogs* (May 11). http://blogs.plos.org/globalhealth/2016/05/is-an-unprecedented-infant-feeding-transition-underway/

Baker, Phillip, Julie Patricia Smith, Libby Salmon, Sharon Friel, George Kent, Alessandro Iellemo, Jai Prakash Dadhich, and Mary J. Renfrew). "Global trends and patterns of commercial milk-based formula sales: is an unprecedented infant and young child feeding transition underway?" *Public Health Nutrition*. May. No. 1. doi:10.1017/S1368980016001117 http://papers.ssrn.com/sol3/papers.cfm?abstract_id=2786419

BALLE. 2016. BALLE: Business Alliance for Local Living Economies. Website. https://bealocalist.org/about-us

Banik, Dan. 2011. "Poverty, Inequality, and Democracy: Growth and Hunger in India." *Journal of Democracy* 22(3): 90-104.

Barnett, Michael. 2012. *The Empire of Humanity: A History of Humanitarianism.* Ithaca, New York: Cornell University Press.

Bartick, Melissa, and Arnold Reinhold. 2010, "The Burden of Suboptimal Breastfeeding in the United States: A Pediatric Cost Analysis." *Pediatrics* 125(5): e1048-56. http://pediatrics.aappublications.org/content/125/5/e1048.long

Bassi, Daniele. 2014. "From Child Hunger to Obesity: Brazil's New Health Scourge." *The Guardian* (May 19). http://www.theguardian.com/global-development-professionals-network/2014/may/19/brazil-obesity-nutrition-malnutrition

Bayne, Martha. 2013. A Crock-pot of Soup as Community Organizer. *Utne Reader* (March/April). http://www.utne.com/mind-body/community-organizer-zm0z13mazwil.aspx?newsletter=1&utm_content=03.08.13+Mind+and+Body&utm_campaign=2013+ENEWS&utm_source=iPost&utm_medium=email

Bezdomny. 2015. *Santopor Residents Create Commons in Rural Greece Through a DIY Wireless Mesh Network. Shareable Weekly.* http://www.shareable.net/blog/sarantaporo-residents-create-commons-in-rural-greece-through-a-diy-wireless-mesh-network?utm_content=2013-03-29%2004%3A56%3A03&utm_source=VerticalResponse&utm_medium=Email&utm_term=Read%20more%20%26raquo%3B&utm_campaign=DIY%20Wireless%20Network%20in%20Greece%20%26%20Top%20P2P%20Trendscontent

Bhatt, Ela R. 2015. *Anubandh: Building Hundred-Mile Communities.* Ahmedabad, India: Navajivan Publishing house. www.e-shabda.com

References

Bhutta, Zulfiqar A., Jai K. Das, Arjumand Rizvi, Michelle F. Gaffey, Neff Walker, Susan Horton, Patrick Webb, Anna Lartey, and Robert E. Black. 2013. "Evidence-based Interventions for Improvement of Maternal and Child Nutrition: What Can be Done and at What Cost? *The Lancet* 382(9890): 452-77. http://www.thelancet.com/journals/lancet/article/PIIS0140-6736%2813%2960996-4/fulltext

Binswanger-Mkhize, Hans, Jacomina P. De Regt, and Stephen Spector. 2010. *Local and Community Driven Development : Moving to Scale in Theory and Practice.* New Frontiers of Social Policy. World Bank. https://openknowledge.worldbank.org/handle/10986/2418

Black, Robert E., Cesar G.Victora, Susan P. Walker, Zulfiqar A. Bhutta, Parul Christian, Mercedesde Onis, Majid Ezzati, Sally Grantham-McGregor, Joanne Katz, Reynaldo Martorell, Ricardo Uauy, and Maternal and Child Nutrition Study Group. 2013. "Maternal and Child Undernutrition and Overweight in Low-income and Middle-income Countries." *The Lancet.* August. 382(9890):427-51. http://www.thelancet.com/journals/lancet/article/PIIS0140-6736%2813%2960937-X/fulltext?_eventId=login

Blank, Joani. 2014. *Common Meals in Cohousing Communities.* Cohousing: The Cohousing Association of the United States. February 1. http://www.cohousing.org/meals-2001

Blas, Javier. 2009. *Summit Draft Removes Date to End Hunger.* FT.com (Financial Times). (November 11). http://www.ft.com/cms/s/0/93ee2244-ceef-11de-8a4b-00144feabdc0.html

Bohn, Sarah, and Caroline Danielson. 2016. *Income Inequality and the Safety Net in California.* San Francisco: Public Policy Institute of California. http://www.ppic.org/main/publication.asp?i=1190

Bosnich, David A. 1996. "The Principle of Subsidiarity." *Religion & Liberty* 6(July and August). http://www.acton.org/pub/religion-liberty/volume-6-number-4/principle-subsidiarity

Brennan, Emily. 2012. "Trying to Close Orphanages Where Many Aren't Orphans at All." *New York Times* (December 4). http://www.nytimes.com/2012/12/05/world/americas/campaign-in-haiti-to-close-orphanages.html?nl=todaysheadlines&emc=edit_th_20121205&_r=0

Brown, Ajamu. 2013. *Real talk.* http://brooklynmovementcenter.org/post/examining-the-food-justice-color-line/

Building Living Neighborhoods. 2016. Website. http://www.livingneighborhoods.org/ht-0/bln-exp.htm

Burgan, Michael, and Mark Winne. 2012. *Doing Food Policy Councils right: A Guide to Development and Action.* Mark Winne Associates. http://www.markwinne.com/wp-content/uploads/2012/09/FPC-manual.pdf

Butterly, John R., and Jack Shepherd. 2010. *Hunger: The Biology and Politics of Starvation.* Hanover, New Hampshire: Dartmouth College Press.

Cabannes, Yves. 2012. *Pro-poor Legal and Institutional Frameworks for Urban and Peri-urban Agriculture.* Rome: Food and Agriculture Organization of the United Nations. http://www.fao.org/docrep/017/i3021e/i3021e.pdf

California WIC Association. 2006. *The First Defense Against Obesity.* California WIC Association and UC Davis Human Lactation Center. http://calwic.org/storage/documents/wellness/bf_paper1.pdf

Carozza, Paolo G. 2003. "Subsidiarity as a Structural Principle of International Human Rights Law." *American Journal of International Law.* 97:38-79. http://www.asil.org/ajil/carozza.pdf

Case, Anne, and Angus Deaton. 2015. "Rising Morbidity and Mortality in Midlife Among White Non-Hispanic Americans in the 21st century." *Proceedings of the National Academy of Sciences of the United States of America.* 112(9). http://www.pnas.org/content/112/49/15078

Cavanagh, John, and Jerry Mander, eds. 2004. *Alternatives to Economic Globalization: A Better World is Possible. Second Edition*. San Francisco: Berrett-Koehler.

Cave, James. 2013. "6 People Making a Difference in Honolulu." *Honolulu Magazine* (November). http://www.honolulumagazine.com/Honolulu-Magazine/November-2013/6-People-Making-a-Difference-in-Hawaii/index.php?cparticle=1&siarticle=0#artanc

CCare. 2016. Center for Compassion and Altruism Research and Education. Website. http://ccare.stanford.edu/about/mission-vision/

Chapin, Ross. 2012. "10 Ways to Love Where You Live." *Yes Magazine* (June 14). http://www.yesmagazine.org/happiness/step-up-step-out-10-ways-to-love-where-you-live

Charny, Israel. 1994. "Toward a Generic Definition of Genocide." In George Andreopoulos, ed., *The Conceptual and Historical Dimensions of Genocide*. Philadelphia: University of Pennsylvania Press.

Charter Cities. 2012. Website. http://chartercities.org/

Chen, Aimin and Walter J. Rogan. 2004. "Breastfeeding and the Risk of Postneonatal Death in the United States." *Pediatrics*113(5). http://pediatrics.aappublications.org/cgi/reprint/113/5/e435

Chibudgielvr. 2009a. *Rat Loves Cat!* YouTube Video. http://www.youtube.com/watch?v=7ikm3o5hDks

——. 2009b. *Rat Loves Cat! – Rejection (Part 2)*. YouTube Video. http://www.youtube.com/watch?v=B-h2YxyncyU

Chirico, Jennfier, and Gregory S. Farley, eds. 2015. *Thinking Like an Island: Navigating a Sustainable Future for Hawai'i*. Honolulu: University of Hawai'i Press.

Christian, Sena. 2012. "Rising Sea Levels: The View from a Canoe." *Yes! Magazine* (May 3). http://www.yesmagazine.org/issues/9-strategies-to-end-corporate-rule/rising-sea-levels-the-view-from-a-canoe

Collins, Jeffrey. 2012. "College Student's Turtle Project Takes Dark Twist." *ABC News* (December 27). http://abcnews.go.com/US/wireStory/college-students-turtle-project-takes-dark-twist-18076298#.UN4JqLbahgI

Committee on World Food Security. 2016. Website. http://www.fao.org/cfs/en/

Cook, Christopher D. 2015. "Harvesting Profits: The Roots of Our Food Crisis." *The Progressive* (July/August). http://christopherdcook.com/uploads/Harvesting_Profits_The_Progressive_July_2015.pdf

Contri, Doug. 2011. "Empathy and Barriers to Altruism." *Peace and Conflict Review* 6(1). http://www.review.upeace.org/images/pcr6.1.pdf

Courtens, Jean-Paul. 2012. "Blame Industrialized Agriculture, Not Organic Farmers." *Letters to the Editor* (September 13). http://www.letterstotheeditor.com/blame-industrialized-agriculture-not-organic-farmers/

Crawford, Jeffrey. 2012. *7 True Stories of Animals Rescuing People from Certain Death.* Cracked.com November 4. http://www.cracked.com/article_20054_7-true-stories-animals-rescuing-people-from-certain-death.html

Credit Suisse. 2015. *Global Wealth in 2015: Underlying Trends Remain Positive.* Credit Suisse. https://www.credit-suisse.com/us/en/about-us/research/research-institute/news-and-videos/articles/news-and-expertise/2015/10/en/global-wealth-in-2015-underlying-trends-remain-positive.html

Cunningham, Solveig A., Michael R. Kramer, and K. M. Venkat Narayan. 2014. "Incidence of Childhood Obesity in the United States." *New England Journal of Medicine* 370(January): 430-411. http://www.nejm.org/doi/full/10.1056/NEJMoa1309753

CUNY School of Public Health. 2016. *Eating in East Harlem*. New York: CUNY Graduate School of Public Health and New York City Food Policy Center at Hunter College. http://eatingineastharlem.org/wp-content/uploads/2016/03/CUNY_FullReport_FINAL_print.pdf

Darwin, Charles Robert 1936. *The Origin of Species by Means of Natural Selection; or, The Preservation of Favored Races in the Struggle for Life and the Descent of Man and Selection in Relation to Sex*. New York: The Modern Library.

Davies, Lisa. 2016. *Opinion: I Was Blown Away by the Bond Between Strangers Who Share Breast Milk*. TVNZ News. https://www.tvnz.co.nz/one-news/new-zealand/opinion-blown-away-bond-between-strangers-share-breast-milk

Deaton, Angus. 2013. *The Great Escape: Health, Wealth, and the Origins of Inequality*. Princeton, New Jersey: Princeton University Press.

DeBruyne, Nese F., and Anne Leland. 2015. *American War and Military Operations Casualties: Lists and Statistics*. Washington, D.C.: Congressional Research Service. https://www.fas.org/sgp/crs/natsec/RL32492.pdf

Deelstra, Tjeerd. 1987. " Urban Agriculture and the Metabolism of Cities." *Food and Nutrition Bulletin* 9(2): 5-7. http://www.unu.edu/unupress/food/8f092e/8F092E03.htm

De Schutter, Olivier. 2015. "Don't Let Food Be the Problem: Producing Too Much Food is What Starves the Planet." *Foreign Policy* (2015). http://foreignpolicy.com/2015/07/20/starving-for-answers-food-water-united-nations/

Dessewfy, Tibor, and Ferenc Hammer. 1995. "Poverty in Hungary." In Ferenc Hammer, ed. *Critical choices for Hungary*. Budapest, Hungary: Joint Eastern Europe Center for Democratic Education and Governance.

de Waal, Frans. 2009. *The Age of Empathy: Nature's Lessons for a Kinder Society*. New York: Harmony Books.

Diacon, Richard Clarke, and Silvia Guimaraes. 2005. *Redefining the Commons, Locking in Value Through Community Land Trusts*. Leicestershire, United Kingdom: The Building and Social Housing Foundation.

Dooley, Jim. 2009. "Aloun Farms Owners to Admit to Illegally Importing Farm Laborers." *Honolulu Advertiser* (January 12). http://www.honoluluadvertiser.com/apps/pbcs.dll/article?AID=20101120331

Dregger, Leila. 2016. *Ecovillages Worldwide—Local Solutions for Global Problems*. Fellowship for Intentional Community. (June 11). http://www.ic.org/ecovillages-worldwide-local-dreggersolutions-for-global-problems/

Drèze, Jean, and Amartya Sen. 1989. *Hunger and Public Action*. Oxford: Clarendon Press.

Duda, John. 2016. "The Italian Region Where Co-ops Produce a Third of its GDP." *Yes Magazine*. (July 5). http://www.yesmagazine.org/new-economy/the-italian-place-where-co-ops-drive-the-economy-and-most-people-are-members-20160705?utm_source=YTW&utm_medium=Email&utm_campaign=20160708

Eberlein, Sven. 2012a. "Life Is Easier With Friends Next Door." *Yes! Magazine* (July 16). http://www.yesmagazine.org/issues/making-it-home/life-is-easier-with-friends-next-door

———. 2012b. *Where No City Has Gone Before: San Francisco Will Be World's First Zero-Waste Town by 2020*. AlterNet (April 18). http://www.alternet.org/visions/155039/where_no_city_has_gone_before:_san_francisco_will_be_world%27s_first_zero-waste_town_by_2020/

Edelman, Peter. 2012. *So Rich, So Poor: Why It's So Hard to End Poverty in America*. New York: The New Press.

Edkins, Jenny. 2000. *Whose Hunger? Concepts of Famine, Practices of Aid.* Minneapolis: University of Minnesota Press.

Edwards, Chris. 2016. "Big Brother and the Breast." *Cato at Liberty.* (June 2016).

Ehrenfreund, Max. 2015. "The stunning—and Expanding—Gap in Life Expectancy Between the Rich and the Poor." *Washington Post* (September 18). https://www.washingtonpost.com/news/wonk/wp/2015/09/18/the-government-is-spending-more-to-help-rich-seniors-than-poor-ones/

Eisler, Riane. 2007. *The Real Wealth of Nations: Creating a Caring Economics.* San Francisco: Berrett-Koehler.

Eisler, Riane. 2016. *Whole Systems Change: A Framework & First Steps for Social/Economic Transformation.* The Next Systems Project. http://www.thenextsystem.org/wp-content/uploads/2016/03/NewSystems_RianeEisler.pdf

Emerson, Ralph Waldo. 1841. *Self-Reliance.* Ralph Waldo Emerson – Texts. http://www.emersoncentral.com/selfreliance.htm

Engle, Patrice L., Purnima Menon, and Lawrence Haddad. 1997. *Care and Nutrition: Concepts and Measurement.* Washington, D.C.: International Food Policy Research Institute. https://www.ifpri.org/publication/care-and-nutrition-0

ETC Group. 2009. *Who Will Feed Us? Questions for the Food and Climate Crises* (November). Action Group on Erosion, Technology and Concentration. http://www.etcgroup.org/content/who-will-feed-us

——. 2013. *With Climate Change . . . Who Will Feed Us?* ETC Group. http://www.etcgroup.org/sites/www.etcgroup.org/files/Food%20Poster_Design-Sept042013.pdf

Fairfax NZ News. 2012. "Free School Milk in Classrooms." *Stuff. co.nz.* http://www.stuff.co.nz/national/education/8076081/Free-school-milk-back-in-classrooms

FAO 2005a. *Voluntary Guidelines to Support the Progressive Realization of the Right to Adequate Food in the Context of National Food Security*. Rome: Food and Agriculture Organization of the United Nations. http://www.fao.org/docrep/meeting/009/y9825e/y9825e00.htm

——. 2005b. *Community-based Food and Nutrition Programmes: What Makes Them Successful: A Review and Analysis of Experience*. Rome: Food and Agriculture Organization of the United Nations. ftp://ftp.fao.org/docrep/fao/006/y5030e/y5030e00.pdf

——. 2009. *More People than Ever Are Victims of Hunger*. Rome: Food and Agriculture Organization of the United Nations. http://www.fao.org/fileadmin/user_upload/newsroom/docs/Press%20release%20june-en.pdf

——. 2011. *Food, Agriculture and Cities: Challenges of Food and Nutrition Security, Agriculture and Ecosystem Management in an Urbanizing World*. Rome: Food and Agriculture Organization of the United Nations. http://www.fao.org/fcit/fcit-home/food-for-the-cities-position-paper/en/

——. 2012. *Making Agriculture Work for Nutrition: Prioritizing Country-level Action, Research and Support*. Food and Agriculture Organization of the United Nations. Global Forum on Food Security and Nutrition. http://www.fao.org/fsnforum/forum/discussions/agriculture-for-nutrition

——. 2015a. *Designing Nutrition-Sensitive Agriculture Investments: Checklist and Guidance for Programme Formulation*. Rome: Food and Agriculture Organization of the United Nations. http://www.fao.org/documents/card/en/c/6cd87835-ab0c-46d7-97ba-394d620e9f38/

——. 2015b. *Key Recommendations for Improving Nutrition Through Agriculture and Food Systems*. Rome: Food and Agriculture Organization of the United Nations. http://www.fao.org/3/a-i4922e.pdf

———. 2016a. *The Human Right to Adequate Food.* Website. Rome: Food and Agriculture Organization of the United Nations. http://www.fao.org/righttofood/right-to-food-home/en/

———. 2016b. *Family Farming Knowledge Platform.* Website. Rome: Food and Agriculture Organization of the United Nations. http://www.fao.org/family-farming/background/en/

———. 2016c. *Voices of the Hungry.* Website. Rome: Food and Agriculture Organization of the United Nations. http://www.fao.org/in-action/voices-of-the-hungry/en/#.Vz6VvGYQHeQ

———. 2016d. *Methods for Estimating Comparable Prevalence Rates of Food Insecurity Experienced by Adults Throught the World.* Rome: Food and Agriculture Organization of the United Nations. http://www.fao.org/3/a-i4830e.pdf

FAO, IFAD and WFP 2015. *The State of Food Insecurity in the World 2015: Meeting the International Hunger Targets: Taking Stock of Uneven Progress.* Rome: Food and Agriculture Organization of the United Nations. http://www.fao.org/3/a-i4646e.pdf

Feinberg, Joel. 1980. *Rights, Justice, and the Bounds of Liberty.* Princeton, New Jersey: Princeton University Press.

Felice, Willliam F. 2003. *The Global New Deal: Economic and Social Human Rights in World Politics.* Lanham, Maryland: Rowman & Littlefield.

Fernholz, Tim. 2016. "The World Bank is Eliminating the Term 'Developing Country' from its Data Vocabulary." *Quartz* (May 17). http://qz.com/685626/the-world-bank-is-eliminating-the-term-developing-country-from-its-data-vocabulary/

FEWS NET. 2016. Famine Early Warning Systems Network. Website. http://www.fews.net/

FIAN International. 2016. "Indian Tea Workers, A Life Without Dignity." *FIAN Newsletter* (May 1). http://www.fian.org/en/news/article/indian_tea_workers_a_life_without_dignity/

Field, Tory, and Beverly Bell. 2013. *Harvesting justice: Transforming Food, Land, and Agricultural Systems in the Americas.* New Orleans and New York: Other Worlds and U.S. Food Sovereignty Alliance. http://www.otherworldsarepossible.org/sites/default/files/documents/Harvesting%20Justice-Transforming%20Food%20Land%20Ag_0.pdf

Findhorn Ecovillage. 2016. Website. http://www.ecovillagefindhorn.com/index.php

Finley, Ron. 2013. Ron Finley: *A Guerrilla Gardener in South Central LA.* Ted Conversation. http://www.ted.com/talks/ron_finley_a_guerilla_gardener_in_south_central_la.html

Fiorella, Kathryn J., Rona L. Chen, Erin M. Milner, and Lia C. H. Fernald. 2016. "Agricultural Interventions for Improved Nutrition: A Review of Likelihood and Environmental Dimensions." *Global Food Security* 8: 9-47. http://www.sciencedirect.com/science/article/pii/S2211912415300146

Fisher, Max. 2012. "Study: Global Crop Production Shows Some Signs of Stagnating." *Washington Post* (December 24). http://www.washingtonpost.com/blogs/worldviews/wp/2012/12/24/is-production-of-key-global-crops-stagnating/

Fletcher, Michael A. 2015. "Income Inequality has Squeezed the Middle Class Out of the Majority." *Washington Post* (December 9). https://www.washingtonpost.com/news/wonk/wp/2015/12/09/income-inequality-has-squeezed-the-middle-class-out-of-the-majority/

Food Commons. 2016. Website. http://www.thefoodcommons.org/

Food First. 2016. "The True Extent of Hunger: What the FAO Isn't Telling You." Food First Backgrounder. 22(2A). http://foodfirst.org/wp-content/uploads/2016/06/Summer2016Backgrounder.pdf

Food Policy Networks. 2016. Johns Hopkins Center for a Livable Future. Website. http://www.foodpolicynetworks.org/

References

Foundation for Community Encouragement. 2016. Website. http://fce-community.org/

Frater, Jamie. 2010. *Top 10 Cases of Animals Saving Humans.* Listverse. http://listverse.com/2010/03/14/top-10-cases-of-animals-saving-humans/

Free the Slaves. 2016. Website. http://www.freetheslaves.net/about-slavery/slavery-today/

Gallagher, Maureen. 2016. *Intentional Communities: Something Old, Something New.* Fellowship for Intentional Community (February 1). http://www.ic.org/intentional-communities-something-old-something-new/

GAIN. 2016. *mNutrition: Behavior Change in 160 Character or Less?* Secure Nutrition (February 1). https://www.securenutritionplatform.org/Pages/DisplayResources.aspx?RID=400

Galtung, Johan. 2016. "Marinaleda: A Concrete Utopia." *Transcend Media Services* (May 23). https://www.transcend.org/tms/2016/05/marinaleda-a-concrete-utopia/

Gandhi, Maneka. 2013. "Animals Too are Emphathetic." *Mathrubhumni* (March 19). http://www.mathrubhumi.com/english/story.php?id=134370

Gardens for Health International. 2016. Website. http://www.gardensforhealth.org/

Generation Nutrition. 2016. *Nutrition Funding: The Missing Piece of the Puzzle.* http://www.generation-nutrition.org/sites/default/files/editorial/missing_piece_of_the_puzzle.pdf

George, Henry. 1879. *Progress and Poverty: An Inquiry into the Cause of Industrial Depressions and of Want with Increase of Wealth . . . The Remedy.* http://www.wealthandwant.com/HG/PP/toc.htm

Gillespie, Stuart, Judith Hodge, Sivan Yosef, and Rajul Pandya-Lorch eds. 2016. *Nourishing Millions: Stories of Change in Nutrition.* Washington, D.C.: International Food Policy Research Institute. http://ebrary.ifpri.org/utils/getfile/collection/p15738coll2/id/130395/filename/130606.pdf

GEN. 2015. *Global Ecovillage Network: 2015 Report.* https://mail.google.com/_/scs/mail-static/_/js/k=gmail.main.en.sXDiEpUnPe0.O/m=m_i,t/am=nhEPBOCejPuDcQ2jgIz0EQrz3n8-XSo_coAX9SfCR0kVwP_N_h_Ar4FesgUG/rt=h/d=1/rs=AHGWq9AIUlgWTCyr1-lHdARPzFMP-kOsow

Goel, Ritika. 2016. *In This Rajasthan Village, the Right to Food Comes Down to Three Lucky Days a Month.* Scroll.in (June 2). http://scroll.in/article/808503/in-this-rajasthan-village-the-right-to-food-comes-down-to-three-lucky-days-a-month

Goetz, Jennifer L., Dacher Keltner, and Emiliana Simon-Thomas. 2010. "Compassion: An Evolutionary Analysis and Empirical Review." *Psychological Bulletin* 136(3): 351-74. http://greatergood.berkeley.edu/resources/studies#compassion_evolutionary_analysis_empirical_review

Goldsbury, Peter. 2010. *New Tools for Growing Living Organisations and Communities: Rich Lessons from the Whirinaki Rainforest.* Te Whaiti Nui-a-Toi. http://www.tipuake.org.nz/files/pdf/Growing%20Living%20Organisations.pdf

Gopnick, Adam. 2012. *The Table Comes First: Family, France, and the Meaning of Food.* New York: Vintage Books.

Graff, Vincent. 2011. "Eccentric town, Todmorden, growing ALL its own veg." *Mail Online.* http://www.dailymail.co.uk/femail/article-2072383/Eccentric-town-Todmorden-growing-ALL-veg.html

GRAIN 2012. *The Great Food Robbery: How Corporations Control Food, Grab Land, and Destroy the Culture.* https://www.grain.org/article/entries/4501-the-great-food-robbery-a-new-book-from-grain.pdf

GRAIN. 2014. *How Does the Gates Foundation Spend its Money to Feed the World.* (November 4). https://www.grain.org/article/entries/5064-how-does-the-gates-foundation-spend-its-money-to-feed-the-world

Grant, Adam. M. 2013a "Does Studying Economics Breed Greed?" *Huffington Post: The Blog* (October 22). http://www.huffingtonpost.com/adam-grant/does-studying-economics-b_b_4141384.html

Grant, Adam M. 2013b. *Give and Take: A Revolutionary Approach to Success.* New York: Viking/Penguin

Grayson, Jennifer. 2016 *Unlatched: The Evolution of Breastfeeding and the Making of a Controversy.* New York: Harper.

Grassroots Institute for Fundraising Training. 2016. Website. http://www.grassrootsfundraising.org/

Green, Marcus Harrison. 2016. "In a Tiny House Village, Portland's Homeless Find Dignity." *Yes! Magazine* (January 28). http://www.yesmagazine.org/peace-justice/-in-a-tiny-house-village-portlands-homeless-find-dignity-20160128?utm_source=YTW&utm_medium=Email&utm_campaign=20160129

Greenspan, Alan. 1966. "Gold and Economic Freedom." Originally published in Ayn Rand's *Objectivist* newsletter in 1966, and republished in her *Capitalism: The Unknown Ideal* in 1967. http://www.constitution.org/mon/greenspan_gold.htm

Griffin, G. Edward. 2010. *The Creature of Jekyll Island: A Second Look at the Federal Reserve.* Fifth edition. Westlake Village, California: American Media.

Guo, Jeff. 2016. "The Sinister, Secret History of a Food that Everybody Loves." *Washington Post* (April 25). https://www.washingtonpost.com/news/wonk/wp/2016/04/25/the-secret-ancient-history-of-the-potato-that-could-change-the-story-of-civilization/?wpmm=1&wpisrc=nl_p1wemost-partner-1

Gupta, Arun. 2013. "Breastfeeding Needs More Aid." *Devex* (August 1). https://www.devex.com/en/news/breastfeeding-needs-more-aid/81555?source=DefaultHomepage_Center_1

Hancox, Dan. 2013. *The Village Against the World: A Communist Utopia in Marinaleda, Spain* (October). https://placesjournal.org/article/the-village-against-the-world/?gclid=CLDwgrK58cwCFQFsfgod8UwDsA

Harmony with Nature. 2016. Website. www.harmonywithnatureun.org/index.html

Hayes, Shannon. 2013. "Instead of Trying to Feed the World, Let's Help it Feed Itself." *Yes! Magazine* (February 20). http://www.yesmagazine.org/blogs/shannon-hayes/instead-trying-feed-world-lets-help-it-feed-itself?utm_source=wkly20130222&utm_medium=email&utm_campaign=mrHayes

Helphand, Kenneth I. 2006. *Defiant Gardens: Making Gardens in Wartime*. San Antonio, Texas: Trinity University Press.

Heifer International. 2013. *Passing on the Gift*. Heifer International http://www.heifer.org/ourwork/approach/passing-on-the-gift

Herforth, Ann. 2012. *Synthesis of Guiding Principles on Agriculture Programming for Nutrition*. https://www.securenutritionplatform.org/Documents/Synthesis%20of%20Ag-Nutr%20Guidance_Sept%202012.pdf

Herman, Louis G. 2013. *Future Primal: How Our Wilderness Origins Show Us the Way Forward*. Novato, California: New World Library.

Hickel, Jason. 2016. "The True Extent of Global Poverty and Hunger: Questioning the Good News Narrative of the Millennium Development Goals." *Third World Quarterly* 37(5).

Hines, Colin. 2004. *A Global Look to the Local: Replacing Economic Globalization with Democratic Localization*. London: International Institute for Environment and Development. http://pubs.iied.org/9308IIED.html?k=colin%20hines

References

HMBNA. 2016. Human Milk Banking Association of North America. Website. https://www.hmbana.org/

Hoffman, Stanley. 1981. *Duties Beyond Borders.* New York: Syracuse University Press.

Howard, Ebenezer. 1902. *Garden Cities of Tomorrow.* http://www.library.cornell.edu/Reps/DOCS/howard.htm

Human Rights Council. 2010. *Report Submitted by the Special Rapporteur on the Right to Food, Olivier De Schutter.* United Nations General Assembly. December 20. A/HRC/16/49. http://www.srfood.org/images/stories/pdf/officialreports/20110308_a-hrc-16-49_agroecology_en.pdf

IFPRI 2013. *2012 Global Food Policy Report.* Washington, D.C.: International Food Policy Research Institute. http://www.ifpri.org/publication/2012-global-food-policy-report?utm_source=New+At+IFPRI&utm_campaign=08cdc30745-New_at_IFPRI_Mar_19_2013&utm_medium=email

Illich, Ivan. 1973. *Tools for Conviviality.* Marion Boyars Publishers. http://www.mom.arq.ufmg.br/mom/arq_interface/3a_aula/illich_tools_for_conviviality.pdf

Immokalee. 2008. Website of Coalition of Immokalee Workers. http://www.ciw-online.org/

Incredible Edible Todmorden Unlimited. 2013. Website http://www.incredible-edible-todmorden.co.uk

Intentional Communities. 2016. Website. http://www.ic.org/

International Food Policy Research Institute. 2016. *Global Nutrition Report 2016: From Promise to Impact: Ending Malnutrition by 2030.* Washington, D.C.: IFPRI. http://bit.ly/GNReport2016

International Human Rights Clinic. 2013. *Nourishing Change: Fulfilling the Right to Food in the United States*. New York: New York University School of Law. http://chrgj.org/wp-content/uploads/2013/05/130527_Nourishing-Change.pdf

International Physicians for the Prevention of Nuclear War, Physicians for Social Responsibility, and Physicians for Global Survival. 2016. *Body Count: Casualty Figures after 10 Years of the "War on Terror.* http://www.psr.org/assets/pdfs/body-count.pdf

International Year of Cooperatives. 2016. Website. http://usa2012.coop/home

Inter Pares. 2004. *Community-based Food Security Systems: Local Solutions for Ending Chronic Hunger and Promoting Rural Development*. Inter Pares: Ottawa, Canada. http://www.interpares.ca/en/publications/pdf/food_security_brief.pdf

Intervale Center. 2016. Website. http://www.intervale.org/

Ioby 2016.Website. http://ioby.org/about

IPES Food. 2016. *From Uniformity to Diversity: A Paradigm Shift from Industrial Agriculture to Diversified Agroecological Systems*. International Panel of Experts on Sustainable Food Systems. www.ipes-food.org

Ipsos-MORI. 2007. *What Works in Community Cohesion*. London: Department for Communities and Local Government. http://resources.cohesioninstitute.org.uk/Publications/Documents/Document/DownloadDocumentsFile.aspx?recordId=72&file=PDF version

Itkowitz, Colby. 2016. "Harvard Researchers Discovered the One Thing Everyone Needs for Happier, Healthier Lives." *Washington Post* (March 2). https://www.washingtonpost.com/news/inspired-life/wp/2016/03/02/harvard-researchers-discovered-the-one-thing-everyone-needs-for-happier-healthier-lives/?wpmm=1&wpisrc=nl_p1most-partner-1

References

IUF. 2009. "Ethical? Tetley's Tata Tea Starving Indian Tea Workers into Submission." *IUF Uniting Food, Farm and Hotel Workers World-Wide* (November 12). http://www.iuf.org/cgi-bin/dbman/db.cgi?db =default&uid=default&ID=6316&view_records=1&ww=1&en=1

Jackson, Will. 2015. "Local Breast Milk for Sale in the US." *The Phnom Penh Post* (December 24). http://www.phnompenhpost.com/national/local-breast-milk-sale-us

Jennings, Katherine. 2013. "Fresh Moves." *Utne Reader* (November 27). http://www.utne.com/food/fresh-moves.aspx?newsletter=1&utm_source=Sailthru&utm_medium=email&utm_term=UTR%20eNews&utm_campaign=12.02.13%20Utne%20eNews#axzz2mLFll4Ws

Joburg. 2013/Caring Cities. 2013. Website. http://joburg2013.metropolis.org/

Johnson, Eric Michael. 2013. "Survival of the . . .Nicest? Check Out the Other Theory of Evolution." *Yes! Magazine* (May 3). http://www.yesmagazine.org/issues/how-cooperatives-are-driving-the-new-economy/survival-of-the-nicest-the-other-theory-of-evolution?utm_source=ytw20130503&utm_medium=email

Jones, Andrew D., and Gebisa Ejeta, 2016. "A New Global Agenda for Nutriton and Health: The Importance of Agriculture and Food Systems." *Bulletin of the World Health Organization* 94(3). http://www.who.int/bulletin/volumes/94/3/15-164509/en/

Jonsson, Urban. 2015. "Faltering Steps." *World Nutrition* 6(9-10): 711-17.

Kanahele, George Hu'eu Sanford. 1986. *Kū Kanaka: Stand tall: A Search for Hawaiian Values*. Honolulu: University of Hawai'i Press and Waiaha Foundation.

Katzenberger, Al. no date. *A Synopsis of Henry George's Progress and Poverty*. Wealth and Want. http://www.wealthandwant.com/HG/PP/Katzenberger_synopsis.html

Kaufman, Frederick. 2009. "Let Them Eat Cash: Can Bill Gates Turn Hunger Into Profit?" *Harper's Magazine*. (June). http://harpers.org/archive/2009/06/let-them-eat-cash/

Kaufman, Frederick. 2012. *Bet the Farm: How Food Stopped Being Food*. New York: Wiley.

Kelemen, Deborah. 2012. "Teleological Minds: How Natural Intuitions about Agency and Purpose Influence Learning about Evolution." In *Evolution Challenges: Integrating Research and Practice in Teaching and Learning about Evolution*, edited by K. S. Rosengren, S. K. Brem, E. M. Evans, and G. M. Sinatra. Oxford: Oxford University Press. http://www.bu.edu/cdl/files/2013/08/2012_Kelemen.pdf

Kelemen, Deborah, Joshua Rottman, and Rebecca Seston 2012. "Professional Physical Scientists Display Tenacious Teleological Tendencies: Purpose-based Reasoning as a Cognitive Default." *Journal of Experimental Psychology: General* (October 15). http://www.bu.edu/childcognition/publications/2012_KelemenRottmanSeston.pdf

Kelly, Marjorie. 2012. "Can There Be 'Good' Corporations?" *Yes! Magazine* (April 16). http://www.yesmagazine.org/issues/9-strategies-to-end-corporate-rule/can-there-be-201cgood201d-corporations?utm_source=april12&utm_medium=email&utm_campaign=mrGoodCorporations

Keltner, Dacher, and Jonathan Haidt. 2003. "Approaching Awe, A Moral, Spiritual, and Aesthetic Emotion." *Cognition and Emotion*. 17(2): 297-314. http://greatergood.berkeley.edu/dacherkeltner/docs/keltner.haidt.awe.2003.pdf

Kennedy, Brendan. 2012. "I Am the River and the River is Me: The Implications of a River Receiving Personhood Status." *Cultural Survival Quarterly* (November 26). http://www.culturalsurvival.org/publications/cultural-survival-quarterly/i-am-river-and-river-me-implications-river-receiving

Kent, George. 1981. "Community-Based Development Planning." *International Development Planning Review* (formerly *Third World Planning Review*) 3(3): 313-326. http://www2.hawaii.edu/~kent/cbdp.pdf

——. 1982. "Food Trade: The Poor Feed the Rich," *Food and Nutrition Bulletin* 4(4): 25-33 http://www.unu.edu/Unupress/food/8F044e/8F044E05.htm

——. 1984. *The Political Economy of Hunger: The Silent Holocaust.* New York: Praeger Publishers.

——. 1986. "Aquaculture: Motivating Production for Low-Income Markets." *Ceres: FAO Review on Agriculture and Development* 19(4): 23-27.

——. 1988. "Nutrition Education as an Instrument of Empowerment." *Journal of Nutrition Education* 20(4): 193-5. http://www2.hawaii.edu/~kent/NutEdGK.pdf

——. 1990. "The Children's Holocaust." *Internet on the Holocaust and Genocide* (Jerusalem) (28(September): 3-6. http://www2.hawaii.edu/~kent/ChildrensHolocaust.pdf

——. 1993a. "The Denial of Children's Mortality." *Internet on the Holocaust and Genocide*, Special Triple Issue 44-46(September): 18, 20. http://www2.hawaii.edu/~kent/DenialofChildrens.pdf

——. 1993b. "Valuation in development projects: Enlarging the analytical framework." *Futures*, 25(8): 902-6. http://www2.hawaii.edu/~kent/Valuation%20in%20Development%20Projects.pdf)

——. 1994. "The Massive Mortality of Children." In *The Widening Circle of Genocide* edited by Israel W. Charny, 272-89. New Brunswick, New Jersey: Transaction Publishers. http://www2.hawaii.edu/~kent/MassiveMortalityofChildren.pdf

——. 1999. "Children's Mortality and Genocide." In *Encyclopedia of Genocide*, Volume I, edited by Israel W. Charny, 138-41. Santa Barbara, California: ABC-CLIO. http://www2.hawaii.edu/~kent/CHILDRENS%20MORTALITY%20AND%20GENOCIDE.pdf

—. 2005. *Freedom from Want: The Human Right to Adequate Food.* Washington, D.C.: Georgetown University Press. http://press.georgetown.edu/sites/default/files/978-1-58901-055-0%20w%20CC%20license.pdf

—. 2006a. "Children as Victims of Structural Violence." *Societies Without Borders.* 1(1): 49-63. http://www2.hawaii.edu/~kent/ChildrenAsVictims.pdf

—. 2006b. "WIC's Promotion of Infant Formula in the United States." *International Breastfeeding Journal.* 1(8).

—. 2008a. *Designing a World Without Hunger.* Transcend Research Institute. 6(August 1). https://www.transcend.org/tri/downloads/Designing%20a%20World%20Without%20Hunger.pdf

—, ed. 2008b. *Global Obligations for the Right to Food.* New York: Rowman & Littlefield.

—. 2009. *The Humiliation of Hunger.* Human Dignity and Humiliation Studies. http://www.humiliationstudies.org/documents/KentHawaii09meetingHunger.pdf

—. 2010. "The Hunger Holocaust." *Genocide Prevention Now.* 2 (April 30). http://www.genocidepreventionnow.org/GPNSearchResults/tabid/64/ctl/DisplayArticle/mid/400/aid/120/Default.aspx

—. 2011a. *Ending Hunger Worldwide.* Boulder, Colorado: Paradigm Publishers.

—. 2011b. *Regulating Infant Formula.* Amarillo Texas: Hale Publishing.

—. 2012a. "The Nutritional Adequacy of Infant Formula." *Clinical Lactation* 3(1): 21-5. http://www.ingentaconnect.com/content/springer/clac/2012/00000003/00000001/art00004

―. 2012b. "ICDS: Steering an Ungainly Ship." *Economic & Political Weekly* (Mumbai, India). XLVII(37): 32-4. http://www2.hawaii.edu/~kent/ICDS_Steering_an_Ungainly_Ship.pdf

―. 2013. "Rights and Obligations in International Humanitarian Assistance." In Peter Bobrowsky, ed., *Encyclopedia of Natural Hazards*. Heidelberg, Germany: Springer. http://www2.hawaii.edu/~kent/RightsandObligationsinIHA24Nov09.doc

―. 2014a. "Building Nutritional Self-reliance." In *Improving Diets and Nutrition: Food-based Approaches,* edited by Brian Thompson and Leslie Amoroso, 268-81. Rome, Italy: Food and Agriculture Organization of the United Nations. http://www2.hawaii.edu/~kent/BuildingNutritionalSelfReliance.pdf

―. 2014b. "Regulating Fatty Acids in Infant Formula: Critical Assessment of U.S. Policies and Practices." *International Breastfeeding Journal* 9(2). http://www.internationalbreastfeedingjournal.com/content/9/1/2

―. 2014c "Regulating the Nutritional Adequacy of Infant Formula in the United States." *Clinical Lactation*, 5(4): 133-6. http://www.ingentaconnect.com/content/springer/clac/2014/00000005/00000004/art00006

―. 2015a. "Food Systems, Agriculture, Society: How to End Hunger." *World Nutrition* 6(3). http://wphna.org/wp-content/uploads/2015/02/WN-2015-06-03-170-183-George-Kent-How-to-end-hunger.pdf

―. 2015b. "Food Systems, Agriculture, and Society: How to Nourish Society." *World Nutrition* 6(4). http://wphna.org/wp-content/uploads/2015/03/WN-2015-06-04-280-291-George-Kent-How-to-nourish-society.pdf

―. 2015c "Global Infant Formula: Monitoring and Regulating the Impacts to Protect Human Health." *International Breastfeeding Journal* 10(6). http://www.internationalbreastfeedingjournal.com/content/10/1/6/abstract

——. 2015d. "On Caring." In Michelle Brenner, ed., *Conversations on Compassion*. Sydney, Australia: Holistic Practices Beyond Borders. http://www2.hawaii.edu/~kent/OnCaring.pdf

King, Barbara. 2013. *How Animals Grieve*. Chicago: University of Chicago Press.

Kiva 2016. Website. http://www.kiva.org/about

Kluger, Jeffrey. 2013. "The Mystery of Animal Grief." *Time* (April 15).

Kohala Center. 2011. "Laulima Center Offers Assistance for Cooperatives." *Hawaii 24/7* (February 8). http://www.hawaii247.com/2011/02/08/laulima-center-offers-assistance-for-cooperatives/

Kolata, Gina. 2015. "Death Rates Rising for Middle-Aged White Americans, Study Finds." *New York Times*. (November 2). http://www.nytimes.com/2015/11/03/health/death-rates-rising-for-middle-aged-white-americans-study-finds.html

Kraak, Vivica I., Stefanie Vandevijvere, Gary Sacks, Hannah Brinsden, Corinna Hawkes, Simón Barquera, Tim Lobstein, and Boyd A. Swinburn. 2016. "Progress achieved in restricting the marketing of high-fat, sugary and salty food and beverage products to children." *Bulletin of the World Health Organization*. 94:540-8. http://www.who.int/bulletin/volumes/94/7/15-158667.pdf

Kropotkin, Peter. 1902. *Mutual Aid: A Factor of Evolution*. Anarchy Archives. http://dwardmac.pitzer.edu/Anarchist_Archives/kropotkin/mutaidcontents.html

——. 1906. *The Conquest of Bread*. New York: G. P. Putnam's Sons. http://dwardmac.pitzer.edu/Anarchist_Archives/kropotkin/conquest/toc.html

Krznaric, Roman. 2012. "The Power of Introspection." Video. *RSA Animate*. http://www.youtube.com/watch?feature=player_embedded&v=BG46IwVfSu8

———. 2013. "6 Habits of Highly Empathetic People." *Yes Magazine* (January 10). http://www.yesmagazine.org/happiness/6-habits-of-highly-empathetic-people?utm_source=wkly20130111&utm_medium=email&utm_campaign=mrKrznaric

Kuhnlein, Harriet V., Bill Erasmus, and Dina Spigelski, eds. 2009. *Indigenous Peoples' Food Systems: The Many Dimensions of Culture, Diversity and Environment for Nutrition and Health*. Rome: Food and Agriculture Organization of the United Nations and McGill University, Canada: Centre for Indigenous Peoples' Nutrition and Environment. http://www.fao.org/docrep/012/i0370e/i0370e00.htm

Künnemann, Rolf. 2016. *Twelve Policies How States Can Make Good Use of Extraterritorial Human Rights Obligations*. Heidelberg, Germany: FIAN International.

Labonte, Ronald, Ted Schrecker, David Sanders, and Wilma Meeus. 2004. *Fatal Indifference: The G8, Africa and Global Health*. Landsdowne, South Africa: University of Cape Town Press. http://www.idrc.ca/en/ev-45682-201-1-DO_TOPIC.html

Laenui, Poka. 2013. "From the Culture of Aloha, a Path Out of Gun Violence." *Yes!* (February 7). http://www.yesmagazine.org/issues/how-cooperatives-are-driving-the-new-economy/violence-guns-and-deep-cultures?utm_source=wkly20130208&utm_medium=email&utm_campaign=mrLaenui

Lampman, Jane. 2000. "Moral Darwinism: The Fittest Conscience: A New Take on Evolution." *Christian Science Monitor*. (August 3). http://www.101bananas.com/library2/lampman.html

Lappé, Frances Moore 2011a. *EcoMind: Changing the Way We Think, To Create the World We Want*. New York: Nation Books.

———. 2011b. "The Food Movement: Its Power and Possibilities." *The Nation*. (September 14). http://www.thenation.com/article/food-movement-its-power-and-possibilities/

—. 2016. *Farming for a Small Planet: Agroecology Now.* Great Transition Initiative. (April). http://www.greattransition.org/publication/farming-for-a-small-planet

Leimbach, Dulcie. 2016. "Universal Breast-Feeding Goals Blocked by Industry, UN Report Says." *Huffington Post.* http://www.huffingtonpost.com/dulcie-leimbach/universal-breastfeeding-g_b_10459112.html

Leong, Lavonne. 2012. "Peer-to-Peer Potential: Building a Culture of Empathy at Punahou." *Punahou Bulletin* (Winter). http://www.punahou.edu/page.cfm?p=3865

Letchworth Garden City. 2016. Website. http://www.letchworth.com/

Levitt, Emily J., David L. Pelletier, and Alice N. Pell 2009. "Revisiting the UNICEF Malnutrition Framework to Foster Agriculture and Health Sector Collaboration to Reduce Malnutrition: A Comparison of Stakeholder Priorities for Action in Afghanistan." *Food Policy* 34(2): 156-165.

Lewenz, Claude. 2007. *How to Build a Village.* Waiheke Island, Auckland, New Zealand: Village Forum Press and Jackson House Publishing Company.

—. 2011. *VillageTowns: The Next Step.* Auckland, New Zealand: Jackson House Publishing Company.

Lifeweb. 2016. Website. www.sahtouris.com

Lightman, Alan. 2014. "Our Lonely Home in Nature." *New York Times* (May 2). http://www.nytimes.com/2014/05/03/opinion/our-lonely-home-in-nature.html?emc=edit_th_20140503&nl=todaysheadlines&nlid=2155033&_r=0

Lincicome, Scott. 2016. *Promoting Free Trade in Agriculture.* The Heritage Foundation. Backgrounder 3136. http://www.heritage.org/research/reports/2016/07/promoting-free-trade-in-agriculture

References

Lindgren Suzanne. 2013. "Bet the Farm: Spinning Wheat into Gold." *UTNE Reader* (January/February). http://www.utne.com/politics/bet-the-farm-zm0z13jfzlin.aspx?newsletter=1&utm_content=01.02.13+Environment&utm_campaign=2013+ENEWS&utm_source=iPost&utm_medium=email

Litfin, Karen. 2010. "The Sacred and the Profane in the Ecological Politics of Sacrifice." In *The Environmental Politics of Sacrifice*, edited by Michael Maniatis and John Meyer. Cambridge, Massachusetts: MIT Press.

———. 2012. *Seed Communities: Ecovillage Experiments Around the World.* You Tube video. http://www.youtube.com/watch?v=MtNjZaXDGqM

Lindner, Evelin. 2012. *A Dignity Economy: Creating an Economy That Serves Human Dignity and Preservers Our Planet.* Dignity Press.

Lobello, Carmel. 2013. "Why Asia is Letting Millions of Tons of Extra Rice Go To Waste." *This Week* (July 31). http://theweek.com/article/index/247611/why-asian-countries-are-letting-millions-of-tons-of-extra-rice-go-to-waste

Lopez Community Land Trust. 2016. Website. http://www.lopezclt.org/

Loye, David. 2007. *Darwin's Lost Theory: Bridge to a Better World.* Third Edition. Benjamin Franklin Press.

———. 2016. Website www.davidloye.com

Luna, Mira. 2013. "Region in Italy Reaches 30% Coop Economy." *Shareable* (July 25). http://www.shareable.net/blog/illustrious-region-in-italy-reaches-30-coop-economy?utm_content=kent%40hawaii.edu&utm_source=VerticalResponse&utm_medium=Email&utm_term=Read%20more&utm_campaign=Shareable%3A%20Region%20in%20Italy%20Reaches%2030%25%20Coop%20Economycontent

Macaulay, Jacqueline, and Leonard Berkowitz.1970. *Altruism and Helping Behavior*. New York: Academic Press.

Making Caring Common. 2016. Website. http://mcc.gse.harvard.edu/

Malkin, Stacy. 2016. "Bill Gates: Can We Have an Honest Conversation about GMOs?" *Ecologist* (March 8). http://www.theecologist.org/News/news_analysis/2987368/bill_gates_can_we_have_an_honest_conversation_about_gmos.html

Manifesto. 1981. *The Manifesto – Appeal of the Nobel Prizewinners*. http://servizi.radicalparty.org/documents/index.php?func=detail&par=44

Martin-Prével, Alice, and Frédéric Moussea. 2016. *The Unholy Alliance: Five Western Donors Shape a Pro-Corporate Agenda for African Agriculture*. Oakland California: Oakland Institute. http://www.oaklandinstitute.org/sites/oaklandinstitute.org/files/unholy_alliance_web.pdf

Mason, Paul. 2015. "The Strange Case of America's Disappearing Middle Class." *The Guardian* (December 14). http://www.theguardian.com/commentisfree/2015/dec/14/the-strange-case-of-americas-disappearing-middle-class?CMP=share_btn_fb

Mata, Leonardo J. 1978. *The Children of Santa Maria Cauqué: A Prospective Field Study of Health and Growth*. Cambridge, Massachusetts: The MIT Press. http://ccp.ucr.ac.cr/bvp/pdf/salud/mata.pdf

Mattera, Philip. 2013. *Nestlé: Corporate Rap Sheet*. Corporate Research Project. http://www.corp-research.org/nestle

McKibben, Bill. 2008. *Deep Economy: The Wealth of Communities and the Durable Future*. New York: St. Martin's Griffin.

Merriam-Webster. 2016. "Schadenfreude" *Merriam-Webster Online Dictionary*. http://www.merriam-webster.com/dictionary/schadenfreude)

Messer, Ellen, and Marc J. Cohen. 2007. *The Human Right to Food as a U.S. Nutrition Concern, 1976-2006*. International Food Policy Research Institute.

Middelmann-Whitney, Conner. 2010. "Conviviality Now! Family Feasts for Body and Soul." *Psychology Today* (May 1). https://www.psychologytoday.com/blog/nourish/201005/conviviality-now

Milan Urban Food Policy Pact. 2016. Website. http://www.foodpolicymilano.org/en/urban-food-policy-pact-2/

Minchin, Maureen. 2015a. "Excerpt from *Milk Matters: Infant Feeding & Immune Disorder*." http://leadertoday.breastfeedingtoday-llli.org/milk-matters-infant-feeding-and-immune-disorder/

——. 2015b. *Milk Matters: Infant Feeding & Immune Disorder*. Milk Matters Pty Ltd.

Ministry of Health of Brazil. 2014. *Dietary Guidelines for the Brazilian Population*, Second Edition. Brasilia. http://www.foodpolitics.com/wp-content/uploads/Brazilian-Dietary-Guidelines-2014.pdf

Minus, Jeff. 2004. "UN, EU, World Court, Supreme Court: Subsidiarity, Anyone?" *Catholic Culture*.(November 22). http://www.catholicculture.org/highlights/highlights.cfm?ID=38

Monsanto. 2016. *What is Monsanto Doing to Address the Issue of Food Security?* http://www.monsantohawaii.com/with-the-worlds-growing-population-food-security-is-a-major-concern-what-is-monsanto-doing-to-address-this-issue/

Monteiro, Carlos Augusto, and others. 2015. "Dietary Guidelines to Nourish Humanity and the Planet in the Twenty-first Century. A Blueprint for Brazil." *Public Health Nutrition* 18(13): 2311-22. http://journals.cambridge.org/action/displayAbstract?fromPage=online&aid=9940567&fileId=S1368980015002165

Morton, Lois Wright, Ella Annette Bitto, Mary Jane Oakland, and Mary Sand. 2008. "Accessing Food Resources: Rural and Urban Patterns of Giving and Getting Food." *Agriculture and Human Values* 25: 107-19. http://link.springer.com/article/10.1007%2Fs10460-007-9095-8#/page-1

Mulvany, Patrick, and Jonathan Ensor. 2011. "Changing a Dysfunctional Food System: Towards Ecological Food Provision in the Framework of Food Sovereignty." *Food Chain* 1(1): 34-51.

Mustain, Patrick. 2013. "Dear American Consumers: Please Don't Start Eating Healthfully. Sincerely, the Food Industry." *Scientific American* (May 19). http://blogs.scientificamerican.com/guest-blog/dear-american-consumers-please-dont-start-eating-healthfully-sincerely-the-food-industry/

Nair, Chandran. 2016. "There's a Place in India Where Religions Coexist Beautifully and Gender Equality is Unmatched." *The World Post/Huffington Post* (April 6). http://www.huffingtonpost.com/chandran-nair/kerala-religion-gender_b_9577234.html

National Center for Children in Poverty. 2016. Website. http://www.nccp.org/topics/childpoverty.html

nef. 2016. National Accounts of Well-Being. Website. http://www.nationalaccountsofwellbeing.org/

New York State Department of Labor. 2016. *Occupational Wages*. Website. https://labor.ny.gov/stats/lswage2.asp#45-0000

NICE 2010. *Donor Milk Banks: The Operation of Donor Milk Bank Services*. National Institute for Health Care and Excellence. United Kingdom. http://www.nice.org.uk/guidance/CG93

Nonhuman Rights Project. 2016. Website. http://www.nonhumanrightsproject.org/

Nohara, Yoshiaki and Andy Sharp. 2013. "Silver Shoplifters Steal Bowls of Rice as Abe Cuts Welfare." *Bloomberg* (July 15). http://www.bloomberg.com/news/2013-07-15/silver-shoplifters-steal-rice-as-abe-cuts-welfare-to-trim-debt.html

Nowak, Martin A. 2012. "Why We Help." *Scientific American* (July): 34-39. http://ped.fas.harvard.edu/files/ped/files/sciam12_0.pdf

Nowak, Martin A., and Roger Highfield. 2012. *Super Cooperators: Altruism, Evolution, and Why We Need Each Other to Succeed.* Free Press.

Oakland Institute. 2015. *Agroecology Case Studies.* Oakland California: Oakland Institute. http://www.oaklandinstitute.org/agroecology-case-studies

On the Hill. 2009. 'Not Whether We Can End Hunger, It's Whether We Will'; Aid Group Endorses Admin. Plan (September 27). http://onthehillblog.blogspot.com/2009/09/not-whether-we-can-end-hunger-its.html

Open Source Ecology. 2016. Website. http://opensourceecology.org/

Ostrom, Elinor. 1990. *Governing the Commons: The Evolution of Institutions for Collective Action.* Cambridge, United Kingdom: Cambridge University Press.

— . 2012. "Green from the Grassroots." *Project Syndicate.* http://www.project-syndicate.org/commentary/green-from-the-grassroots Also available through *Yes! Magazine* at http://www.yesmagazine.org/planet/ostrom-rio-20?utm_source=wkly20120622&utm_medium=email&utm_campaign=titleOstrom

Other Worlds. 2013. Website. http://www.otherworldsarepossible.org/home-page

OXFAM. 2016. *An Economy for the 1%.* OXFAM Briefing Paper (January 18). https://www.oxfam.org/sites/www.oxfam.org/files/file_attachments/bp210-economy-one-percent-tax-havens-180116-en_0.pdf

Oxfam America. 2016. *No Relief: Denial of Bathroom Breaks in the Poultry Industry.* Oxfam America. https://www.oxfamamerica.org/static/media/files/No_Relief_Embargo.pdf

Partnership for Maternal, Newborn & Child Health. 2012. *Born Too Soon: The Global Action Report on Preterm Birth.* New York (May 2). http://www.who.int/pmnch/media/news/2012/preterm_birth_report/en/

Pascual, Tara, and Jessica Powers. 2012. *Cooking up Community: Nutrition Education in Emergency Food Programs.* WHY Hunger and National Hunger Clearinghouse. http://www.whyhunger.org/uploads/fileAssets/CUCFINAL1.pdf)

PATH. 2013. *Strengthening Human Milk Banking: A Global Implementation Framework, Version 1.1.*Seattle, Washington, USA: Bill & Melinda Gates Foundation Grand Challenges Initiative. http://www.path.org/publications/detail.php?i=2433 http://www.path.org/publications/files/MCHN_strengthen_hmb_frame_April2015.pdf

Pattern Language. 2016. Website. http://www.patternlanguage.com/

Patwari, Ashok K., Sanjay Kumar, and Jennifer Beard. 2015. "Undernutrition Among Infants Less Than 6 Months of Age: An Underestimated Public Health Problem in India." *Maternal and Child Nutrition* 11: 119-126. http://onlinelibrary.wiley.com/doi/10.1111/mcn.12030/epdf

Pew Research Center. 2015. *The American Middle Class is Losing Ground: No Longer the Majority and Falling Behind Financially.* Washington, D.C: Pew Research Center. http://www.pewsocialtrends.org/2015/12/09/the-american-middle-class-is-losing-ground/

Phillips, Tom. 2007. "Brazil's Ethanol Slaves: 200,000 Migrant Sugar Cutters Who Prop Up Renewable Energy Boom." *The Guardian* (March 9). http://environment.guardian.co.uk/energy/story/0,,2030144,00.html

phys.org. 2013. "Can Plants Be Altruistic? You Bet, Study Says." phys.org (February 1). http://phys.org/news/2013-02-altruistic.html

Piketty, Thomas. 2014. *Capital in the Twenty-First Century*. Cambridge, Massachusetts: Harvard University Press.

Pilisuk, Marc, and Susan Hillier Parks. 1986. *The Healing Web: Social Networks and Human Survival*. Hanover, New Hampshire.

Pimbert, Michel. 2009. *Towards Food Sovereignty: Reclaiming Autonomous Food Systems*. London: International Institute for Environment and Development. http://www.iied.org/towards-food-sovereignty-reclaiming-autonomous-food-systems

Pogge, Thomas. 2004. "The First United Nations Millennium Development Goal: A Cause for Celebration?" *Journal of Human Development* 5(3). http://biblioteca.hegoa.efaber.net/registro/ebook/14570/The_First_United_Nations_Millennium_Development_Goal.pdf

———. 2016. "The Hunger Games." *Food Ethics*. 1:9-27. http://link.springer.com/article/10.1007/s41055-016-0006-9

Polanyi, Karl. 1944. *The Great Transformation*. Boston, Massachusetts: Beacon Press. http://inctpped.ie.ufrj.br/spiderweb/pdf_4/Great_Transformation.pdf

Pole, Antoinette, and Margaret Gray. 2013. "Farming Alone? What's up with the 'C' in Community Supported Agriculture." *Agriculture and Human Values* 30: 85-100. https://www.researchgate.net/publication/257511421_Farming_alone_What's_up_with_the_C_in_community_supported_agriculture

Pollan, Michael. 2007. *The Omnivore's Dilemma: A Natural History of Four Meals*. New York: Penguin.

———. 2013. "The Intelligent Plant." *The New Yorker* (December 26 and 30): 92-105. http://www.newyorker.com/reporting/2013/12/23/131223fa_fact_pollan?currentPage=all

Pope Francis. 2016. *Show Mercy to our Common Home*. http://w2.vatican.va/content/francesco/en/messages/pont-messages/2016/documents/papa-francesco_20160901_messaggio-giornata-cura-creato.html

Popova, Maria. 2013. "Givers, Takers, and Matchers: The Surprising Science of Success." *Brain Pickings*. http://www.brainpickings.org/index.php/2013/04/10/adam-grant-give-and-take/

PS21 Project. 2015. *Death Toll in 2014's Bloodiest Wars Sharply Up on Previous Year*. PS21 Project for the Study of the 21st Century. https://projects21.files.wordpress.com/2015/03/ps21-conflict-trends.pdf

Putnam, Robert D. 2016. *Our Kids: The American Dream in Crisis*. New York: Simon & Schuster.

Putnam, Robert D., and Lewis M. Feldstein. 2003. *Better Together: Restoring the American Community*. New York: Simon & Schuster,

Rasmussen Derek. 2013. "'Non-Indigenous Culture': Implications of a Historical Anomaly." *Yes!* (July 11). http://www.yesmagazine.org/peace-justice/non-indigenous-culture-implications-of-a-historical-anomaly

Ratner, Carl. 2012. *Cooperation, Community, and Co-Ops in a Global Era*. Springer.

REconomy Project. 2016. Website. http://www.reconomyproject.org/

Robbins, Liz, and N. R. Kleinfield. 2012. "4 Firefighters Shot, 2 Fatally, in New York; Gunman Dead." *New York Times* (December 24). http://www.nytimes.com/2012/12/25/nyregion/2-firefighters-killed-in-western-new-york.html?pagewanted=all&_r=0

Robi, Terri. 2014. *U.S. Explanation of Position on the Right to Food*. HumanRights.gov. http://usun.state.gov/remarks/6295 and http://www.humanrights.gov/dyn/2014/12/u/u.s.-explanation-of-position-on-right-to-food

Rodrigue, Edward P., and Richard V. Reeves. 2014. *Nursing Opportunity: Class Gaps in Breastfeeding and Policy Challenges* (September 10). http://www.brookings.edu/blogs/social-mobility-memos/posts/2014/09/10-nursing-opportunity-class-gaps-in-breastfeeding-reeves

Rose, Donald, J., Nicholas Bodor, and Mariana Chilton. 2006. "Has the WIC Incentive to Formula-Feed Led to an Increase in Overweight Children?" *Journal of Nutrition* 136(4): 1086-90. http://jn.nutrition.org/content/136/4/1086.long

Rosenthal, Elisabeth. 2013. "As Biofuel Demand Grows, So Do Guatemala's Hunger Pangs." *New York Times* (January 5). http://www.nytimes.com/2013/01/06/science/earth/in-fields-and-markets-guatemalans-feel-squeeze-of-biofuel-demand.html?_r=0

Roser, Max. 2016a. *Hunger and Undernourishment.* Our World in Data. https://ourworldindata.org/hunger-and-undernourishment/

Roser, Max. 2016b. *War and Peace after 1945.* Our World in Data. https://ourworldindata.org/war-and-peace-after-1945/

Ross, Carne. 2012. *Guide to the Leaderless Revolution: How Ordinary People Will Take Power and Change Politics in the 21st Century.* Penguin/Blue Rider Press. http://www.carneross.com/sites/www.carneross.com/files/A_Brief_Guide_to_the_LR.pdf

Ross, Janell. 2016. "3 Charts that Challenge the Conventional Wisdom of 2015." *Washington Post* (January 2). https://www.washingtonpost.com/news/the-fix/wp/2016/01/02/3-charts-that-challenge-the-political-conventional-wisdom-of-2015/?wpmm=1&wpisrc=nl_p1wemost

Rowland, Lyndal. 2016. *Breastfeeding Saves Lives But Can't Compete with Aggressive Marketing.* Inter Press Service (May 11). http://www.ipsnews.net/2016/05/breastfeeding-saves-lives-but-cant-compete-with-agressive-marketing/

RSF Social Finance. 2016. Website. http://rsfsocialfinance.org/services/

RTFN-Watch. 2012. *The Right to Food and Nutrition Watch*. http://www.rtfn-watch.org/

Sacks, Jonathan. 2012. "The Moral Animal." *New York Times* (December 23). http://www.nytimes.com/2012/12/24/opinion/the-moral-animal.html?nl=todaysheadlines&emc=edit_th_20121224

Sahlins, Marshall. 1972. *Stone Age Economics*. Chicago: Aldine-Ahterton. https://libcom.org/files/Sahlins%20-%20Stone%20Age%20Economics.pdf

Sahtouris, Elisabet. 1998. *The Biology of Globalization*. http://www.ratical.org/LifeWeb/Articles/globalize.html

Sang-Hun, Choe. 2013. "As Families Change, Korea's Elderly Are Turning to Suicide." *New York Times* (February 16). http://www.nytimes.com/2013/02/17/world/asia/in-korea-changes-in-society-and-family-dynamics-drive-rise-in-elderly-suicides.html?ref=choesanghun&_r=0

Santa Barbara, Jack, Fred Dubee, and Johan Galtung. 2009. *Peace Business*. Oslo, Norway: Kolofon Press.

Sanz-Cañada, Javier. 2016. "Local Agro-Food Systems in America and Europe. Territorial Anchorage and Local Governance of Identity-based Foods." Culture & History Digital Journal, 5(1): e001 http://cultureandhistory.revistas.csic.es/index.php/cultureandhistory/article/view/88/307

Savory Institute. 2016. Website. http://savory.global/institute

Scaling Up Nutrition. 2016. Website. http://scalingupnutrition.org/

Scheiber, Noam, and Patricia Cohen. 2015. "For the Wealthiest, a Private Tax System That Saves Them Billions." *New York Times* (December 29). http://www.nytimes.com/2015/12/30/business/economy/for-the-wealthiest-private-tax-system-saves-them-billions.html?emc=edit_na_20151229&nlid=2155033&ref=cta&_r=0

Schukoske, Jane E. 1999. "Community Development Through Gardening: State and Local Policies Transforming Urban Open Space." *Legislation and Public Policy* 3(35): 351–92. https://communitygarden.org/wp-content/uploads/2013/12/schukoske.pdf

Schulman, Mark. 2012. "How to Combat a Culture of Violence—and Maybe Save Lives." *Huffington Post: The Blog* (December 27). http://www.huffingtonpost.com/mark-schulman/combat-culture-of-violence_b_2371661.html

Schumacher, Ernst Friedrich. 1973. *Small is Beautiful: Economics as if People Mattered*. Blond & Briggs, HarperCollins. http://www.ditext.com/schumacher/small/small.html or http://www.ditext.com/schumacher/small/small.html

Schwartz, Nelson D. 2016. "In an Age of Privilege, Not Everyone Is in the Same Boat." *New York Times* (April 23). http://www.nytimes.com/2016/04/24/business/economy/velvet-rope-economy.html?emc=edit_th_20160424&nl=todaysheadlines&nl id=2155033&_r=0

SCN. 2013. *SCN News. Special Issue on Changing Food Systems for Better Nutrition*. Geneva: United Nations System Standing Committee on Nutrition. No. 40. http://www.unscn.org/files/Publications/SCN_News/SCNNEWS40_final_standard_res.pdf

Shareable. 2016. *How is Technology Changing the Way We Share Food? The SHARECITY Research Team to Find Out*. Shareable. http://www.shareable.net/blog/how-is-technology-changing-the-way-we-share-food-the-sharecity-research-team-to-find-out?utm_content=2013-03-29%20 04%253A56%253A03&utm_source=VerticalResponse&utm_medium=Email&utm_term=Read%20more%20 %2526raquo%253B&utm_campaign=Crowdsourced%20Book%20 Proposal%252C%20Sharing%20Cities%20Chat%20%2526%20 Toy%20Librariescontent

Simoes, Eduardo, and Inae Riveras. 2008. Amnesty Condemns Forced Cane Labor in Brazil. Reuters (May 29). http://in.reuters.com/article/environmentNews/idINN2844873820080528?sp=true

Simon, Michele. 2012. "Feds to Parents: Big Food Still Exploiting Your Children; Good Luck With That." *Food Safety News* (January 8). http://www.foodsafetynews.com/2013/01/feds-to-parents-big-food-still-exploiting-your-children-good-luck-with-that/#.UO59q7bahgI

Singer, Tania, and Matthieu Ricard, eds. 2015. *Caring Economics: Conversations on Altruism and Compassion, Between Scientists,* Economists, and the Dalai Lama. New York: Picador.

Slow Money. 2016. Website. http://www.slowmoney.org/

Solar Gardens 2016. *Solar Gardens: Community Power.* Website. http://www.solargardens.org/

Starling, Shane. 2016. *Whistle Stop Tour: Protein Global.* Nutraingredients.com (March 24). http://www.nutraingredients.com/Markets-and-Trends/Whistle-stop-tour-Protein-global/?utm_source=Newsletter_Subject&utm_medium=email&utm_campaign=Newsletter%2BSubject&c=QyPOfLNUzfOt7MxCxaH4JRzRDQPgWqgV

Steinberg, Scott. 2012. *The Crowdfunding Bible: How to Raise Money for Any Startup, Video Game, or Project.* Readme. http://www.crowdfundingguides.com/

Steinberger, Jillian Laurel, and Scott Peterson. 2016. *The Seed Saving Controversy.* Edible Feast. http://www.ediblefeast.com/eddyawards/vote/9358/seed-saving-controversy

Sterken, Elisabeth. 2016. *Protections for Breastfeeding at the 69th World Health Assembly.* Lactation Matters: The ILCA Blog. http://lactationmatters.org/

Stiglitz, Joseph E. 2013. *The Price of Inequality: How Today's Divided Society Endangers our Future.* New York: W. W. Norton.

References

Stockholm Resilience Center. 2016. http://www.stockholmresilience.org/21/research/research-news/1-31-2013-rediscovering-the-maya-way.html

Stone, Christopher D. 1974. *Should Trees Have Standing: Toward Legal Rights for Natural Objects*. Los Altos, California: William Kaufmann, 1974.

———. 1987. *Earth and Other Ethics: The Case for Moral Pluralism* New York: Harper & Row.

Sustainable Food Trust 2016. *Panel Discussion: Food Justice*. YouTube video. https://www.youtube.com/watch?v=ZW_i4GBBvc0

Sutton-Redner, Jane. 2014. "Once a Dust Bowl, Now a Place of Plenty." *World Vision Magazine* (August). https://magazine.worldvision.org/stories/once-a-dust-bowl-now-a-place-of-plenty

Syngenta. 2016. *The Good Growth Plan*. Website. http://www.syngenta.com/global/corporate/en/goodgrowthplan/home/Pages/homepage.aspx

Swinburn, Boyd 2016. "Health Minister Sets Shameful Legacy." *New Zealand Herald*. (June 10). http://m.nzherald.co.nz/nz/news/article.cfm?c_id=1&objectid=11653688

Tavernise, Sabrina. 2015. "Disparity in Life Spans of the Rich and the Poor is Growing." *New York Times* (February 12). http://www.nytimes.com/2016/02/13/health/disparity-in-life-spans-of-the-rich-and-the-poor-is-growing.html?emc=edit_th_20160213&nl=todayshe adlines&nlid=2155033

The Biodynamic Land Trust. 2016. Website. https://www.biodynamiclandtrust.org.uk/about/what-is-a-land-trust/

The Economist. 2009. "Triple Bottom Line: It Consists of Three Ps: Profit, People and Planet." *The Economist* (November 17). http://www.economist.com/node/14301663

——. 2012. "The Q&A: Mary Robinson. Speaking Truth to Power." *The Economist* (October 3). http://www.economist.com/blogs/prospero/2012/10/qa-mary-robinson)

——. 2016a. "India's Rural Economy: Dry Times." *The Economist* (April 30). http://www.economist.com/news/finance-and-economics/21697870-meagre-rainfall-only-part-problem-indias-farmers-dry-times

——. 2016b. "Biotechnology: Seedy Business." *The Economist* (May 21). http://www.economist.com/news/business/21699142-its-eat-or-be-eaten-firms-make-seeds-and-chemicals-farmers-seedy-business

The Economist Intelligence Unit. 2016. *Global Food Security Index*. Website. http://foodsecurityindex.eiu.com/

The Lancet. 2016. *Lancet Breastfeeding Series* (January 29). http://www.thelancet.com/series/breastfeeding

The Law Dictionary. 2016. "What is Willful Negligence?" http://thelawdictionary.org/willful-negligence/

Thompson, Mark. 2015. "Memorial Day's True Legacy: A More Peaceful World, Amid Continuing War. *Time* (May 29). http://time.com/4350776/memorial-days-true-legacy/?xid=newsletter-brief

Today Pets & Animals. 2010. *Pet Saviors: 11 Animals Who Saved Human Lives*. NBCNews.com (April 28). http://today.msnbc.msn.com/id/36834168/ns/today-today_pets_and_animals/t/pet-saviors-animals-who-saved-human-lives/#.UO55VqWIM7Q

Townsend, Joseph. 1786. *A Dissertation on the Poor Laws*. http://socserv2.socsci.mcmaster.ca/~econ/ugcm/3ll3/townsend/poorlaw.html

Tran, Mark. 2013. "Guatemala Remembers Conflict Victims as New Battles Ignite Over Resources." *The Guardian* (October 24). http://www.theguardian.com/global-development/2013/oct/24/guatemala-battle-resources

Transition Network. 2016. Website. http://www.transitionnetwork.org/

Transnational Institute. 2015. *Investing for Development?* TNI (September 17). https://www.tni.org/en/publication/investing-for-development

———. 2013a. "The Founding Fables of Industrialised Agriculture." *Independent Science News* (October 30). http://www.independentsciencenews.org/un-sustainable-farming/the-founding-fables-of-industrialised-agriculture/

Tsai, Marisa. 2016. *Brazil's Dietary Guidelines: Eat Real Food, Together.* FoodTank. http://foodtank.com/news/2016/07/brazils-dietary-guidelines-eat-real-food-together

Tudge, Colin. 2013b. *Why Genes Are Not Selfish and People Are Nice.* Edinburgh, Scotland: Floris Books.

———. 2013c. "World Agriculture: Living Well Off the Land. [Commentary] *World Nutrition* 4(7): 514-48. http://wphna.org/wp-content/uploads/2015/02/WN-2013-04-06-361-390-Colin-Tudge-Living-well-off-the-land-1.pdf

Ubuntu Age. 2012. *We Have Entered The AGE of UBUNTU.* http://www.harisingh.com/UbuntuAge.htm

UNCTAD. 2013. *Wake Up Before it is Too Late: Make Agriculture Truly Sustainable Now for Food Security in a Changing Climate.* United Nations Conference on Trade and Development. Trade and Environment Review. http://unctad.org/en/PublicationsLibrary/ditcted2012d3_en.pdf

UNECOSOC. 1999. *Substantive Issues Arising in the Implementation of the International Covenant on Economic, Social and Cultural Rights: General Comment 12 (Twentieth Session, 1999) The Right to Adequate Food (art. 11).* Geneva, Switzerland: United Nation. Economic and Social Council. E/C.12/1999/5. http://www.refworld.org/docid/4538838c11.html

UNICEF. 2013. *Breastfeeding on the Worldwide Agenda*. New York: United Nations Childrens Fund. http://www.unicef.org/eapro/breastfeeding_on_worldwide_agenda.pdf

UNICEF-WHO-The World Bank. 2014. *Levels & Trends in Child Malnutrition*. http://www.data.unicef.org/corecode/uploads/document6/uploaded_pdfs/corecode/LevelsandTrendsMalNutrition_Summary_2014_132.pdf

——. 2015. *Joint Child Nutrition Estimates—Levels and Trends*. http://www.who.int/nutgrowthdb/estimates2014/en/

United Nations. General Assembly. 2008. *General Assembly Adopts 52 Resolutions*.... New York: United Nations General Assembly GA/10801. Department of Public Information. December 18. http://www.un.org/News/Press/docs/2008/ga10801.doc.htm

United Nations Millennium Project Task Force on Hunger. 2005. *Halving Hunger: It Can Be Done*. London: Earthscan/Millennium Development Project. http://www.unmillenniumproject.org/reports/tf_hunger.htm Breastmilk Banking and Sharing

United States Department of Agriculture. 2016. *Global Food Security*. Website. http://www.ers.usda.gov/topics/international-markets-trade/global-food-security/questions-answers.aspx#security

UNOHCHR 2010a. *Convention on the Prevention and Punishment of the Crime of Genocide*. Geneva, Switzerland: United Nations Office of the High Commissioner for Human Rights. http://www2.ohchr.org/english/law/genocide.htm

——. 2010b. *United Nations Charter*. Geneva, Switzerland: United Nations Office of the High Commissioner for Human Rights. http://www.un.org/en/documents/charter/index.shtml

——. 2010c. *Universal Declaration of Human Rights*. Geneva, Switzerland: United Nations Office of the High Commissioner for Human Rights. http://www.ohchr.org/EN/UDHR/Pages/Introduction.aspx

Urban Roots. 2012. Website. http://www.urbanrootsamerica.com/urbanrootsamerica.com/Home.html

USAID. 2016a. *Costing Approach Raises Awareness of Malnutrition in Guatemala.* Washington, D.C.: United States Agency for International Development. http://www.fantaproject.org/news-and-events/costing-approach-raises-awareness-malnutrition-guatemala

USAID. 2016b. *Research and Policy Papers.* Washington, D.C.: United States Agency for International Development. https://www.usaid.gov/what-we-do/agriculture-and-food-security/food-assistance/resources/research-and-policy-papers

Vauban. 2016. Website. http://www.vauban.de/info/abstract.html

Veggie Mobile. 2012. Website. http://www.capitalroots.org/programs/veggie/veggie/

Viertel, Josh. 2012 "Beyond Voting with Your Fork: From Enlightened Eating to Movement Building." *Food First Backgrounder* 18(1). http://foodfirst.org/wp-content/uploads/2013/12/BK18_1-2012_Spring_Backgrounder_-_Beyond_Voting_with_your_Fork-_From_Enlightened_Eating_to_Movement_Building.pdf

Village Forum. 2016. Village Towns at the Village Forum. Website. http://www.villageforum.com/

Villines, Sharon. 2014. "Cohousing Meal Programs and Leadership." *Sociocracy: A Deeper Democracy* (August 24). http://www.sociocracy.info/cohousing-meal-programs-and-leadership/

Vyawahare, Malavika. 2013. "India's Battle Against Nutrition Data Deficiency." *International New York Times* (October 24). http://india.blogs.nytimes.com/2013/10/24/indias-battle-against-nutrition-data-deficiency/?emc=edit_tnt_20131024&tntemail0=y

Walsh, Bryan. 2013. "Why We Don't Care About Saving Our Grandchildren From Climate Change." *Time* (October 21). http://science.time.com/2013/10/21/why-we-dont-care-about-saving-our-grandchildren-from-climate-change/?xid=newsletter-weekly

Warhurst, Pam. 2012. "How We Can Eat Our Landscapes." *YouTube*. http://www.ted.com/talks/pam_warhurst_how_we_can_eat_our_landscapes.html?utm_source=newsletter_weekly_2012-08-10&utm_campaign=newsletter_weekly&utm_medium=email

Watson, Lilla. 2016. "If you've come here to help me . . ." https://en.wikipedia.org/wiki/Lilla_Watson

Westra, Laura. 2006. *Environmental Justice and the Rights of Unborn and Future Generations: Law, Environmenal Harm and the Right to Health*. London: Earthscan.

Wiggins, Steve, and Sharada Keats 2013. *Smallholder Agriculture's Contribution to Better Nutrition*. London: Overseas Development Institute. http://www.ajfand.net/Volume13/No3/Reprint-ODI%20Smallholder%20agriculture%E2%80%99s%20contribution%20to%20Nutrition%202013.pdf

Williams, Cecily. 1939. *Milk and Murder*. International Organization of Consumers Unions. http://wphna.org/wp-content/uploads/2014/03/1939_Cicely_Williams_Milk_and_murder.pdf

Williamson, Peter. 2016. *Automated Dairy Farming Gains Momentum*. International Milk Genomics Consortium. May. http://milkgenomics.org/article/automated-dairy-farming-gathers-momentum/?utm_source=Newsletter_May2016&utm_campaign=SPLASHmay2016&utm_medium=email

Winne, Mark. 2016. *Ramen U: Is This the New Meal Plan?* Mark's Food Policy Blog (April 25). http://www.markwinne.com/ramen-u-is-this-the-new-meal-plan/

Wong, Kate. 2012. "Lending a Helping Paw: When Animals Cooperate." Slide Show. *Scientific American* (June 19). http://www.scientificamerican.com/article.cfm?id=cooperation-when-animals-cooperate-lending-helping-paw&WT.mc_id=SA_printmag_2012-07

References

Woodiwiss, Catherine. 2013. "Video demonstrates 'mind blowing' U.S. wealth inequality." *Yes! Magazine* (March 8). http://www.yesmagazine.org/new-economy/video-demonstrates-mind-blowing-wealth-inequality?utm_source=wkly20130308&utm_medium=email&utm_campaign=mrVideo

World Bank. 1986. *World Development Report 1986*. Washington, D.C.: World Bank. https://openknowledge.worldbank.org/bitstream/handle/10986/5969/WDR%201986%20-%20English.pdf?sequence=1

———. 2007. *From agriculture to nutrition: Pathways, Synergies and Outcomes*. Washington, D.C.: World Bank, Agriculture and Rural Development Department.

———. 2015a. *World Development Indicators: Fragile Situations Part 1*. Washington, D.C.: World Bank. http://wdi.worldbank.org/table/5.8

———. 2015b. *World Development Report 2015*. Washington, D.C.: World Bank. http://www.worldbank.org/en/publication/wdr2015

———. 2016. *World Development Report 2016: Digital Dividends*. Washington, D.C.: World Bank. http://www.worldbank.org/en/publication/wdr2016

World Food Programme. 2016. *What Causes Hunger?* Geneva: WFP. https://www.wfp.org/hunger/causes

World Health Organization. 1981. *International Code of Marketing of Breast-milk Substitutes*. Geneva: WHO. http://www.who.int/nutrition/publications/code_english.pdf

———. 2003. *Community-based Strategies for Breastfeeding Promotion and Support in Developing Countries*. Geneva: WHO. http://www.who.int/maternal_child_adolescent/documents/9241591218/en/

———. 2009. *Global Health Risks*. Geneva, Switzerland: WHO. http://www.who.int/healthinfo/global_burden_disease/global_health_risks/en/index.html

———. 2014. *Children: Reducing Mortality*. Geneva: WHO. http://www.who.int/mediacentre/factsheets/fs178/en/#

———. 2016a. *Laws to Protect Breastfeeding Inadequate in Most Countries*. Geneva: WHO. http://www.unicef.org/media/media_91075.html

———. 2016b. *Under-five Mortality*. Geneva: WHO. http://www.who.int/gho/child_health/mortality/mortality_under_five_text/en/

———. 2016c. *WHO Guidelines on Nutrition*. Geneva: WHO. http://www.who.int/publications/guidelines/nutrition/en/

———. 2016d. *Comprehensive Implementation Plan on Maternal, Infant and Young Child Nutrition*. Geneva: WHO. http://www.who.int/nutrition/publications/CIP_document/en/

———. 2016e. *Maternal, Infant and Young Child Nutrition: Report by the Secretariat*. Executive Board. 138th Session. Geneva. World Health Organization. http://apps.who.int/gb/ebwha/pdf_files/EB138/B138_8-en.pdf

World Health Organization, UNICEF, and IBFAN. 2016. *Marketing of Breast-milk Substitutes: National Implementation of the International Code: Status Report 2016*. http://www.who.int/nutrition/publications/infantfeeding/code_report2016/en/

Worldwatch. 2009. "Climate Change Reference Guide and Glossary." In *State of the World 2009: Into a Warming World*. Washington, D.C.: Worldwatch Institute.

Worthy, Kenneth. 2013. *Invisible Nature: Healing the Destructive Divide between People and the Environment*. Amherst, New York: Prometheus Books.

York, Geoffrey. 2008. "Vietnam Farmers Face Paradox of the Paddy." *Globe and Mail* (May 1). http://www.theglobeandmail.com/report-on-business/vietnams-farmers-face-paradox-of-the-paddy/article671742/

Zeleznik, Daniel. 2013. *UN General Assembly Commemorates International Mother Earth Day with Harmony with Nature Dialogue.* Boston, Massachusetts: Global Environmental Governance Project (May 13). http://www.environmentalgovernance.org/event/2013/05/un-general-assembly-commemorates-international-mother-earth-day-with-harmony-with-nature-dialogue/

Zeitlin, Marian, Hossein Ghassemi, and Mohamed Mansour. 1990. *Positive Deviance in Child Nutrition with Emphasis on Psychosocial and Behavioural Aspects and Implications for Development.* WHO/UNICEF Joint Nutrition Support Programme. United Nations University. http://bvs.per.paho.org/texcom/nutricion/posdev.pdf

Praise for *Caring About Hunger*

Caring About Hunger is exactly what Professor George Kent has done his whole professional life, bridging the gap between what UN-FAO and governments do and the work by the NGOs in civil society. Read the book, and start caring!

Johan Galtung, Founder of Transcend International: A Peace Development Environment Network

After decades of research on hunger, George Kent—who was already working as a consultant for the Food and Agriculture Organization of the United Nations in the 1980s—argues that there are no technical obstacles preventing us from ending hunger in the world. He makes a convincing case that the "Hunger Holocaust" is first and foremost about how well or how poorly we treat each other in our communities. He provides shocking figures and rigorous analyses in the first chapters, but also offers a compelling vision of how the issue of hunger can actually be solved through strong caring communities. Far from dreaming up far-fetched utopian schemes, Kent proposes small, feasible, but crucial, shifts of awareness regarding this issue. His preface modestly insists that at the very least there is a need to examine the role of caring in relation to hunger. That discussion has now started, with this fascinating, informative and pleasantly written volume.

Olivier Urbain, Director of the Toda Institute for Global Peace and Policy Research, Senior Research Fellow at the Min-On Music Research Institute, author of Daisaku Ikeda's Philosophy of Peace, and editor of Music and Conflict Transformation

Praise for Caring About Hunger

Hunger expert George Kent, with a global vision and breadth of knowledge, provides a moral context and a menu of feasible solutions that make a food-secure world more desirable for everyone.

Linda Riebel
Author of The Green Foodprint: Food Choices for Healthy People and a Healthy Planet

To solve the world's food problems, we of course need good science. But equally we need sound morality – which means for starters that we need to *care*. There is no shortage of food in the world, says George Kent, yet a billion still go hungry – because we, humanity, just don't care enough.

Colin Tudge, Author,
Founder of the College for Real Farming and Food Culture

It is no surprise that *Caring About Hunger* is being published in the 21st century. As George Kent says in the preface, "There is no technical obstacle to ending hunger." What is lacking is the motivation to care. Modern societies have developed ways of protecting and achieving. We have now the challenge of the 21st century to actualize our other natural capacity, caring. In this book, George Kent takes us on a journey to feel the power of caring about hunger. George has thrown down the gauntlet to all governments and all people to care about hunger. Which of them will face the challenge and harness our nature for good?

Michelle Brenner, Founding Member of Holistic Practices Beyond Borders; Conflict Resolution Consultant, Sydney, Australia

George Kent has devoted much of his adult life to understanding world hunger, exploring its causes, and searching for its cures. He has done this through his teaching, research, and advocacy. While we already have endless documentation of the extent of hunger and malnutrition and on agriculture, nutritional needs, and aid programs, Kent turns our attention to the human aspects of the problem. Compassion for the needy is present but weak, often overwhelmed by motivations that permit exploitation of many people. Hunger allows employers to find cheap labor and get high profits, and consumers find cheap products. Kent's work shows how the human right for food needs to be strengthened to ensure that all people have decent opportunities to provide for themselves. The dignity restored to people by assuring the means to feed their families dignifies the entire human family. Kent's book helps us bring the most fundamental of human needs significantly closer to the center of human attention.

Marc Pilisuk, Professor Emeritus, The University of California, and Faculty, Saybrook University

George Kent's latest book reflects the steady evolution of his understanding of hunger and the misery it brings for millions. He reiterates his view that technology will not overcome it as long as the social and political dimensions of the problem are not addressed. The new angle he explores calls for looking inward, not upward, for solutions. His insight is that in strong communities, no one goes hungry--and he explores what this means.
He helps the reader understand this complex issue, showing:
- That food production in the modern world is mainly for the wealth of large producers, while poor rural households are exploited.
- The need to overcome the forces that create and sustain hunger

by more decisively addressing the roots of the problem.
- Why the main challenge is to promote caring in relation to hunger, overcoming the widespread indifference toward it.
- Why breastfeeding is so central in this caring approach.
- Why tackling the problems of worldwide hunger through charity is futile.

The book ends on an optimistic note, and makes a passionate call for agencies to partner more with local communities in addressing the core issues of hunger.

There is much intelligence and insight to mine in this book. It is a timely wake-up call. Timing is everything in comedy and in scholarship. There is no better timing for this book's publication.

Claudio Schuftan, Long time nutrition activist, People›s Health Movement, Saigon, Vietnam

Caring About Hunger is a book of remarkable scope, a politically rich, ethically challenging and sociologically sound text identifying a major caring deficit of our global society. That deficit allows the sustained hunger holocaust, with millions dying every year of hunger-related causes. Despite the title, Kent's book is more about anti-hunger realpolitik than about tenderness. Hunger continues due to indifference and exploitation by "us (well-nourished readers)" in relation to "them (the malnourished that will never read this book)." It continues because we simply do not care enough about hunger. Distancing ourselves from the others' plight and also from nature meshes with the ethos and praxis of the industrial food system, nicely described in this book. Too often, caring about the other is superseded by an irrational form of rational individualism. The author shows that better agricultural technologies and nutritional knowledge will not get rid of hunger unless caring is included in the policy mix. His simple and powerful proposal, based on his extensive life experience and academic knowledge, identifies caring solutions at the bottom instead of proposing blueprints from the top. It is not a matter of us outsiders helping vulnerable, marginalized people, but facilitating their taking care of themselves, on their own terms. You must not miss this book. This is a must-read for food and nutrition professionals and daily eaters, a delightful piece for those who really care.

Jose Luis Vivero Pol, food and nutrition scholar and anti-hunger activist, Université catholique de Louvain, Belgium

This book is a must-read for everyone who cares about the current state of the world. Hunger is more than simply one of many indignities that humankind has to solve, it is at the core, from where all other problems can be analyzed and unraveled. George Kent masterfully shows that the right to adequate food is not a benevolent gift from elites, but is indivisibly linked to every individual's entitlement to dignity.

Evelin Lindner, Founding President, Human Dignity and Humiliation Studies

This extraordinary book is a cogent plea and a treasure of insight on how to tackle the hunger problem. Starting from the realization that food systems are social as well as technological, the author proposes the development of strong communities, strong in the sense that their people care about each other's well-being and about their environment. He shows the principles for designing caring communities, then broadens the scope to describe caring economics, grounded in partnership or mutual respect systems, rather than domination or top-down control systems. The author rightly says, "Nutrition status depends on the quality of the relationships among the people within the community and also on their relationships with others in and from other communities." Thus he advocates solutions that show respect for local people's dignity, where insiders and outsiders strive toward a more even relationship of learning from each other.

Ulrich Spalthoff, Director of Programs and System Administration, Human Dignity and Humiliation Studies

Caring About Hunger, the latest book by Professor George Kent, synthesizes a lifetime of work on building pathways to a world without hunger and malnutrition. Professor Kent's writing and scholarship shine like the eyes of a beloved, well-nourished, breastfed baby. This book will inspire many people, even those with very limited time and resources, to take well-directed practical action. He writes: *"Any small group could take up the challenge of creating a new caring community. One of the first tasks would be to identify a site and gain control over it. In some cases, governments might be persuaded to provide under-utilized land."* Studying Professor Kent's work has changed my life and given me new hope.

Ronald Johnson, postgraduate student, University of Sydney, Australia.

George Kent has written a little book which should be on everyone's bookshelf. From his vast knowledge of the preventable and ongoing human condition which results in more untold misery than almost any other, gathered during a lifetime's study, George has synthesized the very essence of the cause of world hunger. There is enough food in the world to feed us all, but hunger happens because we don't care enough to end it. The solutions are at once both complex and simple. This book provides much food for thought

Pamela Morrison, International Board Certified Lactation Consultant, Britain

In a world dominated by markets, where resources flow toward power instead of need, what would happen if we cared more about hunger? George Kent guides communities to look inwards—not upwards—for solutions. In his new book, Kent explores the unnecessary nature of hunger in a technologically-advanced age, and makes a well-documented argument against hunger from public health, human rights and development standpoints. He challenges both consumers and power-holders to rethink their roles. Acknowledging the difficulty of solving the hunger problem from the top down, and avoiding utopic tendencies, *Caring About Hunger* champions a more practical local approach to solving the hunger problem: empowering local communities to increase their social cohesion. This book is a must-read for anyone who "cares about" making real strides in reducing hunger in the world.

Brooke Aksnes, Léogâne, Haiti; Executive Editor, Nourish Network; Public Health Nutritionist

In *Caring About Hunger*, George Kent provides us with an eloquent appeal to the conscience of individuals and communities worldwide to care about the most basic of human rights, the right to adequate food. More importantly, he builds on his decades of work analyzing global hunger to construct a detailed strategy to turn that awakened conscience into effective action.

Joel Federman, Chair, Transformative Social Change Department, Saybrook University, California

George Kent

After more than forty years of teaching in the University of Hawaii's Department of Political Science, George Kent retired in 2010 as Professor Emeritus. Currently he serves as an Adjunct Professor with the Department of Peace and Conflict Studies at the University of Sydney in Australia and also with the Transformative Social Change Program at Saybrook University in California. He teaches an online course on the Human Right to Adequate Food for both of them.

Professor Kent has worked with the Food and Agriculture Organization of the United Nations, the United Nations Children's Fund, the World Food Programme and several nongovernmental organizations. He is on the Board of Directors of the International Peace Research Association Foundation. His recent books on food policy issues are *Freedom from Want: The Human Right to Adequate Food*, *Global Obligations for the Right to Food*, *Ending Hunger Worldwide*, and *Regulating Infant Formula*.

www.ingramcontent.com/pod-product-compliance
Lightning Source LLC
Chambersburg PA
CBHW071158240526
45470CB00017B/335